Flexible Food Packaging

Flexible Food Packaging

Questions and Answers

Arthur Hirsch, Ph. D., FAIC

VNR VAN NOSTRAND REINHOLD
New York

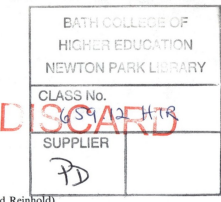

An Avi Book
(AVI is an imprint of Van Nostrand Reinhold)
Copyright © 1991 by Van Nostrand Reinhold

Library of Congress Catalog Card Number 91-17647

ISBN 0-442-00609-8

Manufactured in the United States of America

Published by Van Nostrand Reinhold
115 Fifth Avenue
New York, NY 10003

Chapman and Hall
2–6 Boundary Row
London, SE 1 8HN, England

Thomas Nelson Australia
102 Dodds Street
South Melbourne 3205
Victoria, Australia

Nelson Canada
1120 Birchmount Road
Scarborough, Ontario M1K 5G4, Canada

16 15 14 13 12 11 10 9 8 7 6 5 4 3 2 1

Library of Congress Cataloging-in-Publication Data
Hirsch, Arthur, 1921–
    Flexible food packaging : questions and answers / by Arthur
Hirsch.
        p.   cm.
    Includes index.
    ISBN 0-442-00609-8
    1. Food—Packaging.   2. Plastics in packaging.   I. Title.
TP374.H57   1991
664'.092–dc20   91–17647
                                                      CIP

To JENNIFER
May you continue to ask questions and experience
the joy of finding answers.

Contents

Preface

Packaging is an essential feature of modern life. The science and art of packaging is so vast that no single book or even a multivolume work could hope to cover the entire scope of topics, from Artwork to Zipper.

This volume has selected some of the most commonly raised questions in the field of flexible packaging of food. No claim is made for comprehensive coverage of the field—nor even for an in-depth exploration of a limited number of topics. The novice should find sufficient material here to gain a broad understanding of flexible packaging. The expert's knowledge may be enriched by the case studies and the additional reading lists.

The first topic covered is "Who needs packaging?" We conclude that everyone depends on packaging. Western civilization as we know it today would cease without modern packaging.

The advantages of controlled atmosphere (CA) or modified atmosphere packaging (MAP are reviewed, especially as they apply to the preservation of meat cheese and produce. The need for a moisture and oxygen barrier is analyzed, and materials that provide these properties are presented.

The legal aspects of packaging are confronted—including FDA and USDA oversight, EPA and toxic waste disposal, bar codes, and nutritional labeling. Machinery—especially form-fill-seal (ffs)—is covered in detail, and the influence of the computer on the modern packaging operation is discussed.

Among the food products chosen for detailed consideration are: snacks, frozen foods, fish, fruits and vegetables, bacon, franks, luncheon meats, fresh meat, cheese, and coffee. A section is devoted to special needs of microwaveable food items. The last section offers a glossary of terms to help the reader understand some of the unique vocabulary that has crept into the packaging language.

The content has been illustrated with the aid of 25 figures and enhanced by 41 tables, designed to clarify and document statistical information.

Flexible Food Packaging

Questions and Answers

Q 1: Who Needs Packaging?

A 1: The answer to this question is: EVERYBODY! Let us examine just a few examples of the enormous impact packaging has made on our daily lives.

THE FARMER

Fewer and fewer farmers provide ever increasing quantities of food for an ever expanding population (Table 1-1). Since a large portion of the world has limited growth seasons, the food must be preserved to provide year-round nourishment. Packaging helps bring the crops to market long after the harvest has been completed. If we conceive of packaging as a means for containment, then the farmer has a long history of package utilization. He has brought his products in a variety of containers to the market. His milk was packaged in large cans, his apples in bushels, his corn in baskets, etc. As time and technology progressed, farm products were packaged in convenient retail size. Milk became available in quarts or even smaller containers.

Table 1-1. U.S. Farm Productivity[1]

Year	US Population[2]	Farm Employment	Gross Farm Products[3]
1960	180.7	7.1	N/A
1970	205.1	4.5	55.1
1980	227.8	3.7	148.4
1987	243.9	2.9	153.4

[1]Adopted from *Statistical Abstracts of the US* 1989 (109th ed.), tables 2, 1075, 1094.
[2]Millions.
[3]Millions of dollars.

Fruits and vegetables were canned and could be stored on the shelves for years in the package. In more modern times, the freezer and microwave oven have impacted on the packaging of farm products as well. The heat-in-bag frozen vegetable is old hat by now, but this mode of preserving and serving vegetables is little more than a generation old. Tons of food can be safely stored in this way for prolonged periods. The farmer is consequently assured a year-round market for products which were once seasonal. Because of packaging, the market for farm products, which was local (or regional at best), has now become global. The consumer may occasionally complain about the cost increases of food items or the garbage crisis blamed on packaging materials. However, he must admit that modern packaging of farm products has virtually eliminated the threat of famine from our land.

It is true that modern transportation enables us to fly in fresh fruits, vegetables, and even flowers from various parts of the globe and thus enjoy fresh berries in the middle of the winter. However, the packaged farm product will continue to play an important role in our overall food supply.

THE RETAILER

The modern supermarket depends on prepackaged products for rapid, inexpensive service. Imagine if every customer had to receive personal service—if each item had to be measured or weighed or counted, while the lines of waiting customers grew longer and longer. Under these circumstances our current retailing system would collapse. The cost of this personalized service would have to be passed on to the customer and thus the prices of all retail items would increase very appreciably. Repeated handling of the bulk product for purposes of extracting a small portion would lead to spillage and spoilage. These factors would in turn contribute to diminished quality and increased cost.

Take for example the distribution of milk, a basic food product. Earlier in this century, the farmer would sell his cow's milk to a dairy. There, milk would be pasteurized in 10 gallon cans and delivered to the grocer. The customer would bring his own container—usually a little pot—and have a pint or quart measured from the large container into his bucket. Each time the large can was opened, bacteria would enter and accelerate the spoilage of the milk. Thus the first customer received relatively clean product. However, by the time the twentieth customer had his milk measured out, the product was so badly contaminated as to spoil rapidly. No wonder most would boil their milk under these circumstances. The retailer ran a high risk of spillage as the volume of milk in his can decreased; the container had to be tilted and spillage was thus a real possibility. Also, contaminated product was often poorly refrigerated and would therefore spoil before it was sold. Several quarts of milk could turn sour and be a total loss.

The current retailer works on a very small margin. There are some who believe that the food retailer is ripping off the public. The fact, however, is otherwise. The supermarket chains are working on a minimal profit (usually around 2%). They could not possibly afford large-scale spoilage, nor could they justify the labor cost associated with personalized service. The retailer with the aid of packaging is in a position to deliver to his customer milk of high quality, at a reasonable cost, and at a minimum time interval. Without packaging, the customer would spend a lot of time in lines receiving a poorer quality product, and pay a much higher price for it.

THE CUSTOMER

Packaging has provided the customer with brand identification which assures uniform and reliable quality. The retail establishment taking advantage of prepackaged items, combining premeasurement with prepricing, have made shopping as painless and as rapid as possible. Packaging also provides the consumer easily identifiable storage containers in the home. Whether it is a glass jar of pickles, or a can of peas, or a box of cornflakes, each item can be stored in the pantry and some or all of its contents removed when desired. If these products were brought home wrapped in paper or in a nondescript paper bag, identification of the product would be most difficult if not impossible. Prolonged storage in such wrapping would be undesirable as well.

It is claimed that the depressed status of the Soviet economy can, to a large extent, be blamed on poor packaging. Long lines at the stores are due to outmoded methods of distribution. Each item has to be hand picked and hand wrapped. The consumer spends hours to acquire basic necessities and is thus less inclined to perform to the best of his ability on the job. This system of distribution increases waste and spoilage, causing shortages and outright absence of items off season.

In the U.S. many packaging innovations have been developed with the consumer's needs in mind. "Economy size" was designed to offer the large family lower prices for the bulk purchase. However, more single households have given rise to "portion packs," where the package provides just enough for a one-time use. The package could be ovenable or in some other way utilized to prepare the food for consumption. Many "reclosable" features have been introduced in the last few years. After a portion of the contents has been withdrawn, the package can be reclosed to protect what remains. Tamper resistance has been an issue since several incidents—some fatal— have alerted the consumer to this potential hazard. Food and drugs could be altered if packages were opened and reclosed in the distribution cycle. Several packaging innovations are designed to alert the consumer that packages are no longer in their virgin condition.

INDUSTRY

Many items other than food require packaging in order to be useful in commerce and industry. For example, packaging provides a means of shipping, storing, and displaying a wide array of products from automotive parts, to housewares, clothing, and textiles, to electrical and electronic parts, and many many others. Packaging protects these items from impact, from dust, and other contamination. During shipment it reduces pilferage, it displays the item in the retail store, and provides for easy identification in the stockroom. It is inconceivable that our industrial and commercial establishment could survive if packaging were eliminated.

In short, the question as to who needs packaging has definitely just one answer, and that is EVERYBODY.

ADDITIONAL READING

Anon. 1988. New glossary of packaging terms ready. *Beverage Industry* **79**(4): 11.
Anon. 1987. The leading edge in food packaging. *Packaging Design* **24**(2): 46–50.
Anon. 1987. Study finds resealability tops consumer wish list. *Food and Drug Packaging* **51**(2): 63.
Anthony, S. 1988. Packaging quality and product quality. *Prepared Foods* **157**(3): 97–98.
Ashton, R. 1989. Thirteen other industries buy $13 billion of packaging. *Packaging* **34**(12): 72.
Barrett, R. K. 1988. Talk about convenience. *Canadian Packaging* **41**(2): 25–28.
Burns, R. 1988. Nylon for the nineties. *Plastic Technology* **34**(2): 60–65.
Folkenberg, J. 1987. Beyond the tin can. *FDA Consumer* **23**(9) 34.
Goldberg, J. 1988. Expediting new packaging development from concept to construct. *Good Packaging* **49**(5): 10–11.
Hunter, B. T. 1985. Packaging food. *Consumer Research Magazine* **68**(2): 8–9.
Kreisher, K. R. 1989. 20% annual growth is just a start for coextruded food packaging. *Modern Plastic* **66**(8): 30–33.
Miller, A. 1986. Giving packaging a bad name. *Newsweek* **108**(7): 45.
Sloan, A. E. 1985. Accepting technology (food radiation, aseptic packaging and retort pouch). *Food Engineering* **57**(7): 48 + .

Q 2: Why Controlled or Modified Atmosphere Packaging?

A 2: The package requirements are often determined by the nature of the product to be packaged. If the product is sensitive to atmospheric conditions such as oxygen, carbon dioxide, acidity, or moisture, then it is often desirable to create an artificial environment around the product, which will be maintained within the package in order to avoid changes in the quality or appearance of the product.

Generally, "controlled atmosphere" (CA) is automatically assumed to be low in oxygen. While this is often the case, it is not necessarily the only atmosphere modification desirable. For example, moisture control to prevent corrosion or mold growth might in some instances be more important than presence or absence of oxygen. In all instances, however, it is assumed that the atmosphere provided in the package remains essentially unaltered for the shelf life period of the product. Thus the package must provide an adequate barrier to maintain the specifically designed atmosphere within the package, permitting no component to escape from or to invade the inner sanctum.

A few examples of controlled atmosphere are cited to illustrate the importance of this concept.

OXYGEN

The beneficial biological functions of oxygen are well known. These same biochemical properties may, however, work to the detriment of the packer.

For example, oxygen promotes not only macro but also micro life. Thus, microorganisms will grow within the package provided adequate oxygen, moisture, and nourishment are available. This phenomenon is most prominently found in moist food products, but could also be found in clothing or industrial products which are by nature moist and provide a nutrient such as a starch coating. The exclusion of oxygen may diminish this problem.

Oxygen may also contribute to the deterioration of the product by chemical means. Oxidation of fats and oils leads to rancidity of many an edible product. Oxidation of ferric metals gives rise to rust and corrosion. The chemical action of oxygen on pigments and dyes (normally in the presence of light) is responsible for fade.

Yet there are occasions when high concentrations of oxygen are desirable. For example, in the packaging of fresh meat, the bright red color depends on the chemical reaction between oxygen and myoglobin. According to one theory the meat is best packed in an oxygen enriched environment.

Packaging of various vegetative products are best carried on in a controlled atmosphere containing *some* oxygen. Since produce, fruits, and flowers are living, they do inhale oxygen and exhale carbon dioxide during their life cycle. To package such a product in an atmosphere totally devoid of oxygen would just suffocate it.

MOISTURE

Many products to be packaged are extremely moisture sensitive. It was mentioned earlier that the moisture content of the product may regulate its susceptibility to microbial decay. A favorite method of old, still practiced today, involves drying the product in order to increase its shelf life. The process of moisture removal has been improved in recent years through freeze drying and vacuum drying. One finds a great array of dried products in the supermarket which can be reconstituted by addition of water. In packaging such products the most important factor is the exclusion of moisture. Thus the atmosphere in the package must be as dry as possible.

There are instances when a high moisture content is most desirable. For example, in the packaging of baked products, low moisture content within the package may cause rapid evaporation of moisture from the surface of the product into the air space of the package and thus dry its contents. Consequently the baked goods stale prematurely. Packaging such an item as a gas pack would require the deliberate injection of moisture in order to maintain the relative humidity within the package at a reasonably high level.

CARBON DIOXIDE (CO_2)

This gas is often employed to replace oxygen. CO_2 would certainly inhibit the growth of microorganisms and is thus a suitable replacement atmosphere for protective packaging.

However, carbon dioxide is not totally inert. It reacts with water to form carbonic acid and can in this way actually dissolve in many products which contain high proportions of water. As the gas dissolves in the water, the quantity of gas within the package diminishes and a partial vacuum is actually generated. This transition from a gas to a vacuum package may be detrimental to a semi-rigid package and may bring about the collapse of same. It may also be undesirable in conjunction with a fragile product which could be crushed by the increased pressure.

Since carbon dioxide has the ability to form carbonic acid in the presence of moisture it can impart a flavor to some edible items which will absorb the acid. Not all food products are affected in this manner. Most items that are served hot will expel the carbon dioxide upon heating.

Because of its acidity, carbon dioxide would not be a suitable inert atmosphere for the packaging of metallic products. Corrosion could be a serious consequence of such an atmosphere.

NITROGEN

This gas is a most suitable inert atmosphere for most packaging applications. The gas is abundantly available at relatively low cost, has neither odor nor color and is chemically practically unreactive. The gas, as normally supplied, is extremely dry and attention must be paid to the moisture content of the inert atmosphere, as discussed above.

CASE STUDY 2-1

Fast food chains, such as Burger King and McDonald's, dispense large quantities of lettuce. Refrigeration prolongs the shelf life of the lettuce somewhat, but browning sets in within a few days, creating all sorts of problems for the food chain operation.

It was found that vacuum packaging increased the useful life of the product somewhat. However, after a few days the lettuce suffocated due to lack of oxygen. The taste of the salad deteriorated markedly.

Packaging Unlimited in Atlanta, GA resolved this problem with a breathable packaging material. Shredded lettuce was vacuum packaged in a plastic composite that retained its moisture content but permitted entry of enough oxygen to allow normal respiration to continue. Lettuce survived for 12 days.

CASE STUDY 2-2

Experiments with vacuum packaging of whole fish were very successful. Tests indicated that vacuum did indeed provide substantial benefits in preserving seafood.

A limited field test occasioned several quality complaints. Unfortunately, this led to the erroneous conclusion that controlled atmosphere packaging was unsuitable for fish. It was not the packaging, however, that was at fault. The product selected for this experiment was of such poor sanitary quality as to permit no further preservation by any means (see Q 22).

ADDITIONAL READING

Anon. 1977. Keeping oxygen out keeps color in. *Modern Packaging* **50**(2): 28–29.

Anon. 1977. Vacuum packaged shredded lettuce stays fresh for 12 full days. *Food Engineering* **49**(4): 81–83.

Lingle, R. 1988. CAP for US bakery products: to be or not to be? *Prepared Foods* **157**(3): 91–93.

Park, C. E. et al. 1988. A survey of wet pasta packaged under a CO_2 : N_2 (20 : 80) mixture for staphylococci and their enterotoxins. *Canadian Institute of Food Science and Technology* **21**(1): 109–111.

Peters, J. W. 1974. Packaging film improves shelf life, protects flavor by removing oxygen, halting oxydative changes. *Food Product Development.* **8**(3): 66–68.

Rothwell, T. T. 1988. A review of current European markets for CA/MAP. *Food Engineering* **60**(2): 36–38.

Segura, J. S. 1987. Controlled atmosphere keeps food fresh in Europe. *Packaging* **32**(3): 74–75.

Silmard, R. E. 1988. Seafood and fish. *The National Provisioner* **198**(9): 13–16.

Spaulding, M. 1988. Oxygen absorbers keep food fresher. *Packaging* **33**(1): 8–10.

Q 3: Which Is Better—Vacuum or Gas Packaging?

A 3: The question is not which one of these types is better but rather which one is better suited for a given packaging assignment. Both vacuum and gas packaging do provide a means for increasing the shelf life of the packaged product. The vacuum package affords protection through removal of most of the oxygen surrounding the product. It is erroneous to assume that all the oxygen has been removed from the package. The amount of residual air will depend on several factors: equipment, maintenance of the equipment, the product, the speed of operation, package design, etc.

Each piece of equipment has some inherent limitations. It is best to consult the equipment manufacturer regarding the machinery specification. Can the equipment do the job? One must realize that no machine can run at 100% efficiency. The equipment must be rated appreciably better than the minimum acceptable performance level. It is also well to remember that neglect of maintenance will lead to rapid deterioration of equipment performance. Such equipment needs to be serviced periodically—on a preventive maintenance schedule—and spare parts must be stocked to avoid lengthy shutdowns.

Packaging machinery is normally rated for optimal operating efficiency. Thus a machine rated to package bacon at 90 packages/min. (ppm) should not be operated at 120 ppm. A cycle of about one-third of a second is allowed for evacuation of the package during normal operating speeds. If the speed is increased to 120 ppm, then the evacuation cycle is diminished to less than a quarter of a second. This time interval is inadequate to fully remove the air from the package. The resultant package would appear to be a "leaker" (as if air had seeped back into the evacuated package) and suffer diminished shelf life.

The product may benefit from either vacuum or gas packaging—but not all products are suitable for either of these packaging modes. A soft, fluffy bread or cake would be compressed into a flat, inedible glob in a vacuum package. On the other hand, franks can be given added shelf life by this packaging method. This, however, does not hold true for all franks. Poorly formulated hot dogs contain a large amount of air. In a rapid packaging cycle, the air within the package is evacuated in a fraction of a second. But within a few minutes, the empty space is filled with air which has slowly oozed out of the franks. The package looks like it had never been evacuated. The product as formulated is just unsuitable for vacuum packaging.

Package design may determine the success of its evacuation. An oversized package, for example, requires the evacuation of an excessive volume of air. Tight control on the product size can often accomplish down sizing of the package and result in a better vacuum performance.

DEGREE OF VACUUM

Evacuation—removal of air—may take place to a greater or lesser extent by design. For example, vacuum packaging of angora sweaters can reduce space requirements by more than 50%. However, a very high degree of evacuation is not really required for this particular application. The normal food pack requires a vacuum level of 28″ of mercury (711 mm). Extremely high vacuum levels, demanding an almost absolute 0% of residual oxygen, are rare. However, modern packaging developments have addressed this particular problem with two independent solutions. Higher vacuum levels— or lower residual oxygen—can be obtained by the use of *backup pumps.* This system of placing two (or more) pumps in series on a vacuum line has been employed for some time in the electronics industry. It permits both faster and better evacuation. A still newer approach involves the *scavenger package,* discussed in reply to Q 4.

GAS PACK

The gas package accomplishes the same end as evacuation, replacing the atmosphere with an inert gas. It should be noted, however, that some gas packages retain a considerable amount of residual oxygen. This is attributable to the type of gas packaging machinery employed. In some equipment the gas flushing operation "sweeps" an inert gas through the package— relying on the sweeping action to replace the atmospheric oxygen. This process, especially at high speeds, may leave considerable quantities of oxygen in the package. Product sensitivity and shelf life expectancy must be the guiding factors in the selection of such a gas flush system. Lower residual oxygen tolerance may force packer to select a different gas flushing system. Machinery is available which evacuates the package prior to the introduction of the inert gas. Thus the oxygen level does not depend on the efficiency of the gas sweep but rather on the performance of the evacuation cycle. Such a system can normally deliver a gas-filled package with approximately $\frac{1}{2}\%$ residual oxygen. This compares very favorably to the gas sweep system which often retains 5% oxygen. It is conceivable that the vacuum-gas flush approach could reduce the oxygen level to about 0.01% by introducing a second evacuation cycle, followed by reflushing. This system would be costly and reduce productivity. However, where extremely low oxygen content is essential, this method would certainly provide a means of attaining it.

The nature of the product or package configuration may eliminate vacuum packaging from consideration. For example, whipped cream filled cake cannot be vacuum packed. The vacuum would crush the cake. Even if this calamity could be avoided, the vacuum would collapse the whipped cream, which depends on tiny air bubbles for its consistency.

Nitrogen (N_2) is the gas most frequently employed in the gas flushing operation. This gas is inexpensive, readily available, and above all inert—causing no change in the packaged product. Carbon dioxide (CO_2) is frequently utilized as well, but it must be selected with some caution, as CO_2 is not entirely inert. It will dissolve in water readily and does reduce internal package pressure. This can be an advantage, as in cheese packaging. The absorption of CO_2 actually causes the loose package to tighten up and improves its appearance without any harmful side effects. However, this same gas applied to an alkaline product will lower its pH (make it more acid) and could seriously affect taste, appearance, or functionality of the product. One certainly would not package antacid pills (or fluids) under a blanket of CO_2 (see Q 2).

CASE STUDY 3–1

Packages of frankfurters off the packaging machine appeared "loose." The vacuum level attainable with the equipment seemed suspect. Service personnel suggested a quick and easy check. A small vacuum gauge (approximately 2″ in diameter), purchased from Marsh Instrument Company, was placed into a thermoformed pocket and evacuated and sealed on the packaging machine. The gauge contained in the sealed package indicated actual vacuum level inside the package. The scale reads 0 to 15 psi and can be converted to inches of mercury or millimeters of mercury by readings from

Table 3–1. By employing several gauges one could check out machine dies in a relatively short time. Mechanical problems were quickly corrected.

CASE STUDY 3–2

This case is very similar to the one cited above. However, testing of the equipment revealed no malfunction. Individual packages containing products were placed in a vacuum oven (suitable equipment would be National Vacuum Oven, Fisher Scientific catalog No. 13-261-1 or Isotemp Vacuum Oven, Fisher Scientific catalog No. 13-261-50 or equivalent). Oven must have see-through door and must be able to maintain 760 mm Hg vacuum. The package is placed in the oven. The door is securely latched and evacuation initiated. Evacuation valve is shut at first sign of packaging material lift. If the pressure inside the package is minutely larger than the pressure

Table 3–1.   Air Pressure Conversions

Readings, 0–15 psi scale	Equivalent Vacuum Readings	
	inches mercury	mm mercury
14.7	29.9	760.2
14.5	29.5	749.9
14.25	29.0	736.9
14.0	28.5	724.0
13.75	28.0	711.1
13.5	27.5	698.2
13.25	27.0	685.2
13.0	26.5	672.3
12.75	26.0	659.4
12.5	25.5	646.4
12.25	25.0	633.5
12.0	24.5	620.6
11.75	24.0	607.7
11.5	23.5	594.7
11.25	23.0	581.8
11.0	22.4	568.9
10.0	20.4	517.2
9.0	18.4	465.4
8.0	16.3	413.7
7.0	14.3	362.0
6.0	12.2	310.3
5.0	10.2	258.6
4.0	8.2	206.9
3.0	6.1	155.1
2.0	4.1	103.4
1.0	2.0	51.7

inside the oven chamber, then the package will expand, causing the packaging material to rise. The pressure inside the vacuum chamber, assumed to be equivalent to that inside the package, may be read off the vacuum gauge.

Coincidentally, the problem of this case was traced to product quality. Competitive product packaged on this very same equipment, with the same suspect packaging material, did not develop the same loose appearance. It was thus shown that the problem was neither packaging machinery nor material related.

Customer was advised to seek resolution of his problem in improved sanitation or reformulation of his product.

ADDITIONAL READING

Anon. 1977. Keeping oxygen out keeps color in. *Modern Packaging* **50**(2): 28–29.

Baumgart, J. 1987. Presence and growth of clostridium botulinum in vacuum packaged raw and cooked potatoes. *Chemie Mikrobiologie Technologie der Lebensmittel* **11**(11): 74–80.

Densford, L. 1988. FDA nixes retail use of sous vide processing. *Food and Drug Packaging* **52**(4): 4,40.

Goepfert, J. M. 1988. Comparison of vacuum and modified atmosphere packaging. *The National Provisioner* **198**(11): 13–15.

New, J. H. 1988. Laboratory studies on vacuum and inert gas packaging for control of stored product insects in foodstuffs. *Journal of the Science of Food and Agriculture* **43**(3): 235–244.

Schillinger, U. and Lucke, F. K. 1987. Lactic acid bacteria on vacuum-packed meat and their influence on shelf life. *Die Fleischwirtschaft* **67**(10): 1244–1248.

Terrington, J. 1988. Vacuum in the meat case. *Meat and Poultry* **34**(1): 64–67.

Warmbier, H. C. and Wolf, M. J. 1976. A pouch for oxygen sensitive products. *Modern Packaging* **49**(7): 38–41.

Q 4: Why Is an Oxygen
Barrier Required?

A 4: Many products are sensitive to oxygen. Fats and oils upon exposure to oxygen—especially in the presence of light—undergo chemical changes which lead to flavor alterations referred to as rancidity.

Colors as well as several flavor ingredients of food are subject to changes upon contact with oxygen. Some vitamins are prone to oxidation, as well. Color changes brought about by exposure to oxygen and light in textiles and a variety of other products are universally known as *fade*.

Contact with oxygen should be minimized in order to reduce spoilage caused by bacterial and mold growth. This problem is not entirely restricted to food items. Starch-coated paper and textiles may develop mildew growth. Many other materials are subject to microbial attack. Even as inert a product as gasoline can support bacterial growth.

It has been established that oxygen can spoil the quality of many items of commerce. It is thus best to prevent prolonged exposure of such items during storage and shipment. Packaging of a sensitive product in vacuum or in an inert atmosphere can reduce contact with oxygen. If the packaging material is absolutely impervious to oxygen (such as a tin can) then the product may be preserved indefinitely. However, most packaging materials do not have the enormous barrier of tin plate. Even aluminum foil in thin gauges possesses a few pinholes which permit some oxygen transmission.

The need for oxygen barrier varies not only with the product but also with its normal life expectancy. Table 4–1 shows several packaging materials and their corresponding oxygen transmission rates. It will be noted that the rate of transmission ranges from practically nil, all the way to sev-

Table 4–1. Oxygen Transmission Rate

Material	cc of Oxygen	
	A	B
12 μ polyester/25 μ aluminum/50 μ PE	0.00	0.00
25 μ polyester/PVDC/37 μ PE	0.19	2.95
12 μ polyester/PVDC/50 μ PE	0.61	9.46
10 μ polypropylene/PVDC/50 μ EVA	0.81	12.56
25 μ nylon/50 μ surlyn	2.10	32.55
12 μ polyester/50 μ PE	6.34	98.27
75 μ PE	92.00	1426.00

A. cc O_2/100 sq. in./24 hours/at 23°C
B. cc O_2/sq. m/24 hours/at 23°C

eral hundred ml of oxygen per square meter per 24 hours at a temperature of 23°C.

It should be noted that these transmission rates are reported at about room temperature and normally diminish considerably at lower temperature. Thus if the product is to be kept refrigerated at about 10°C (50°F), the oxygen migration into the package will be appreciably less—possibly less than half the rate reported in Table 4-1.

In most instances the amount of oxygen permissible within the package has never been ascertained. After all, the spoilage of the product depends on a variety of factors of which oxygen is just one component. It is thus practically impossible to state unequivocally that a given product will or will not spoil within a specified period, if kept in an atmosphere containing a certain percentage of oxygen. Experience over many years tells us that in a good vacuum package, processed meat for instance will survive for 45 to 60 days, if properly refrigerated, provided its oxygen barrier is better than 15.5 ml of oxygen per square meter per 24 hours at 23°C (1 cc/100 sq. in/ 24 hours at 73.4° F). One should of course stipulate that the processed meat must be of good quality to start with. Obviously, if the product were of questionable quality, its shelf life would be severely restricted, even with excellent barrier packaging.

A gas barrier is not always necessary or even desirable. If the product is not oxygen sensitive or when the package is neither evacuated nor gas filled, the introduction of a gas barrier lacks justification. Providing a potato chip package with a saran barrier, while leaving normal atmosphere in the package, serves no purpose. Similarly, if one were to gas pack potato chips in a light transparent barrier material, one would severely restrict the benefits available from this costly package. Even minute amounts of oxygen which would penetrate the barrier would set off rancidity in the presence of light.

In the case of frozen food, a gas barrier serves little or no function. Because of the ultralow temperatures, most or all chemical activity has been stopped and oxidation has been practically eliminated. It is therefore not necessary to exclude oxygen from the package. Nevertheless, vacuum packaging is often practiced in the frozen food industry. However, it serves a purpose other than food quality preservation. (For more on this subject see the answer to Q 21.)

A new packaging approach to low oxygen packaging can be found in the patent literature—the scavenger package. It has been mentioned earlier (Q 3) that the normal vacuum package provides for about 0.5% residual oxygen at best. In case this minimal residue cannot be tolerated, the scavenger package claims to reduce the oxygen content to virtually zero. The patent provides for a suspension of palladium and hydrogen in the adhesive system joining the several layers of the packaging laminate. Oxygen inside the package migrates through the sealing layer and reacts with the hydrogen to form water. This process continues until all of the residual oxygen has been consumed. Similarly, no new oxygen is permitted to reach the product, since all oxygen migrating through the outer layer is converted to water on its way into the package. Whether this actually happens is subject to conjecture. The package cost is unknown but assumed to be high. The efficacy of this scavenger, especially over prolonged storage intervals is unknown. FDA attitude toward potential palladium contamination of packaged food remains to be explored.

DETERMINATION

The rate of oxygen transmission through the walls of the package is measurable by various means. It should be remembered that the rate varies with both temperature and humidity. Thus tests should be run at environmental conditions resembling those of actual package storage or use. Furthermore, transmission rates are usually determined for flat sheets of materials. It is important to realize that most packages are not flat and many involve thermoformed composites. The forming process not only thins materials, but may actually disrupt barrier layers. The data gathered from testing of flat sheets may not be applicable to thermoformed packages. Finally, the transmission rate is normally expressed as ml per square meter or cc per 100 square inches. However, the package size may not approach either of these measurements and must be considered (see Case Study 4–1).

TEST PROCEDURES

A. Linde cell. Follow ASTM D1434. The procedure is cumbersome and time consuming. Very accurate results, but unsuitable for other than flat sheets.

B. Ox-tran®. This test instrument, manufactured by Modern Controls, Minneapolis, MN, will measure gas transmission rates (GTR) through packaging materials in 20–55 minutes. The Ox-tran® uses an oxygen-specific fuel cell as a detector to provide a sensitivity down to the parts per billion level, even in the presence of water vapor or other gases. No pressure differential is required across the barrier; consequently, there are no pressure seals, pumps, or vacuum systems to maintain. Gas bubblers provide a means of conditioning the various gases to a relative humidity from 0 to 100% RH, as selected. A temperature control unit attached to the diffusion cell permits studies of the effects of change in GTR with temperature. Test results on barriers with known oxygen transmission rates directly correspond to data obtained using ASTM D-1434 procedures. A permanent record of the oxygen transmission rate is provided by a 0-1-500 millivolt recorder. The continuous graph recording gives an instantaneous visual verification when the barrier has reached equilibrium (Figure 4-1). Temperature controls are available to run temperature profiles in a matter of a few hours. The Ox-tran® is well suited to run oxygen transmission rates on actual packages, not just flat sheet.

CASE STUDY 4-1

A packer of peanuts wanted to provide longer shelf life through the exclusion of oxygen from his package. Through experiments with foil packaging,

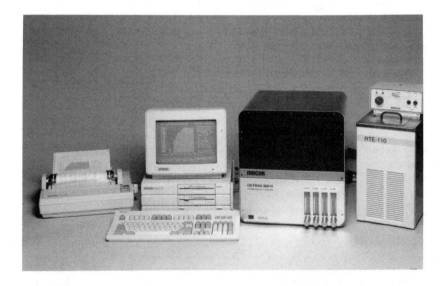

Figure 4-1.   Ox-tron 300 H. (*Courtesy Mocon/Modern Controls, Inc.*)

the packer determined that a pack containing less than 5% oxygen retained the quality of his product for one year.

What is the permissible oxygen transmission rate for a transparent flexible material to yield the same shelf life? The package size was 1.5" × 3.5" and the retained oxygen at time of vacuum packaging was 0.5%.

Answer: Packaging material   = 1.5" × 3.5" × 2 = 10.5 sq. in.
Package volume   = 1 cu. in. = 16.5 ml
Max. O_2 transmission ≅ 4.5% in 365 days = .75 cc/365
                              = .75/365 = 0.002 cc/24 hrs/10.5 sq. in.
                              = 0.002 × 100/10.5
                              = 0.019 cc/100 sq. in./24 hrs.

This calculation excludes practically all but foil composites from consideration.

CASE STUDY 4-2

A foil composite has an oxygen transmission rate of virtually zero. Yet a packer complained of rapid deterioration of his packaged product.

Investigation showed that flexing of the package produced tiny pinholes and consequently diminished barrier properties appreciably. A better understanding of the actual conditions in use might have avoided the problem entirely.

The difficulty was resolved by substituting an extrusion lamination for the prior adhesive laminate. The polyethylene surrounding the aluminum foil provided a cushion which reduced pinholing and cracking drastically. (See also the answers to Q 6 and Q 7, dealing with foil pinholes and metallized polyester.)

ADDITIONAL READING

Anon. 1988. Oxygen absorbing longlife military technology could lead to fresher, longer-lasting foods and beverages. *Good Packaging* 49(4): 22.

Blethen, C. S. 1975. Total test time considerations for complete package gas transmission rate measurement. *Package Development* 5(2): 18–22.

Cramer, G. M. 1987. FDA views on high barrier packaging. *Activity Reports of the R&D Associates* 39(1): 55–58.

Hall, C. W. 1973 Permeability of plastics. *Modern Packaging* 46(11): 53–57.

Herlitzer, W. et al. 1973. Calculating a plastic package for oxygen-sensitive food. *Verpackungs Rundschau* 24(7): 51–55.

Herrero, F. M. and Waibel, J. 1976. Efficient method of investigating gas permeation through packaging materials. *Verpackungs Rundschau* **27**(3): 19–26.

Karel, M. 1974. Packaging protection for oxygen-sensitive products. *Food Technology* **28**(8): 50–65.

Rizvi, S. S. and Gylys, R. B. 1983. Nondestructive method measures air in pouches. *Packaging* **28**(9): 65.

Q 5: When Is a Moisture Barrier Desired?

A 5: Far too often we think of a barrier as a shield that keeps intruders out. The walls of a castle were designed to ward off invaders. Yet the bars on a prison cell are equally construed as barriers to keep the incarcerated in.

In packaging, too, the barrier has this dual function of prohibiting entry into and escape from the interior of the package. In the case of the oxygen barrier, which has been discussed earlier, the primary function is that of excluding the gas from the package content. However, in the case of the moisture barrier, the packaging material may be required to maintain the moisture content at either a low or high level.

Dried foods maintain their quality due to the absence of moisture. Microbial life cannot exist in the absence of water. Under perfectly dry conditions it is easy to store food products for long periods of time without the aid of vacuum or refrigeration. It is, however, most urgent to exclude even minute amounts of moisture, since these would lead to the rapid deterioration of the food item thus exposed. The packaging must exclude the migration of even traces of moisture over the expected storage life of the product. Many chemical and pharmaceutical preparations are hygroscopic: they will absorb moisture, like a sponge, from the air. Often, this absorption will continue until the solid product has caked or has turned into a liquid. Table salt is an example of a product which cakes up. A salt that claims to "pour when it rains" is adulterated with other chemicals which will preferentially absorb water, leaving the salt in a relatively dry state. Bakery products are especially sensitive to moisture content. A properly functioning pack would be designed to maintain the moisture content at the desired level. Similarly, ketchup or mustard would cake up if their moisture concentration were allowed to diminish.

recap To recap then, each food item packaged must be maintained at an optimum moisture level.

The same considerations that were invoked for food packaging are equally applicable to many other types of packages. The potential ill effects of excessive moisture on chemicals and pharmaceuticals were mentioned above. Many other industrial items are sensitive to moisture loss or gain. Paper products will embrittle and/or curl when moisture content varies. Many metallic products are subject to corrosion in the presence of excessive moisture. Cigars are best stored in a humidor to maintain a proper moisture content. Wooden products warp when exposed to a very dry atmosphere for prolonged periods of time. One may go on reciting the influence of variations of moisture content on synthetic fibers, flowers, leather, etc. Proper packaging of items of commerce and industry can preserve such products from either excessive loss or gain of moisture and thus make them impervious to changing atmospheric humidity conditions.

MVTR DETERMINATION

A number of methods are available for testing the moisture vapor transmission rate (MVTR) for packaging materials. Brickman introduced the pouch method in the early 1960s. The pouch was prepared from the actual packaging material to be tested and was filled with either the commercial product or a stand in. Pouches can be stored at various temperature/humidity conditions and weight changes noted at set intervals. If weight changes are plotted, as in Figure 5-1, it becomes possible to predict shelf life stability of the product in the proposed package. Extrapolations are subject to misinterpretation. This test presupposes that the package under test is in every respect identical to its commercial offspring and that test conditions (temperature, humidity, etc.) reflect shipping and storage conditions. The Brickman test has gained even greater importance today. The MVTR of a flat pouch or packaging material can be easily determined by machines introduced in the late 1960s and early 1970s. However, thermoformed packages could not be handled by these electronic devices until much later. The Brickman method could evaluate thermoformed packages as if they were pouches. Actual product or calcium chloride could be sealed into the thermoformed package and these could be handled in the manner of the flat pouch discussed above.

The Honeywell Model W 825 was a giant step forward in MVTR determinations. The instrument eliminated the need for numerous weighings. A

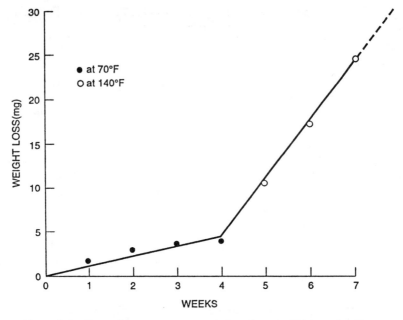

Figure 5-1. Rate of water vapor transmission through .004 in. polyethlyene.

of packaging material was clamped into place above a water reservoir and the rate of water vapor penetration was electronically tested and results recorded. It normally takes several hours to reach equilibrium conditions. A newer instrument, employing an infrared detection system and providing data print out is the Mocon Permatran®. It gives fast results—often within a few minutes (Tables 5-1 and 5-2).

Table 5-1. **Moisture Vapor Transmission Rate (40°C, 90% RH) for Single-Ply Packaging Materials.**

Film	g/25 μ/m^2/24 hrs	g/mil/100 in^2/24 hrs
BOPP	5.9	0.38
HDPE	5.9	0.38
PP	10.7	0.69
LDPE	17.7	1.14
PET	20.2	1.3
PVC	46.5	3.0
PVDC	0.93	0.06
Barex 210	94.6	6.1
MXD6 Nylon	50	3.2

Table 5-2. Water Vapor Transmission Rates (WVTR) for Select Packaging Composites.

Materials	WVTR	
	A	B
18 μ PET/10 μ Al foil/ heat seal coating	0.002	0.031
37 μ Aclar/125 μ Meliform	0.018	0.28
150 μ PVC/37 μ EVA	0.17	2.64
12. μ PET/PVDC/50 μ PE	0.24	3.72
25 μ nylon/50 μ Surlyn	0.96	14.88

PET is polyethylene terephthalate polyester.
A. g/100 sq. in./24 hours.
B. g/m^2/24 hours.

CASE STUDY 5-1

A study was undertaken for a foil producer to determine the moisture penetration through "defective" foil composites. It is well known that thin foil (less than 50 μ or 2 mil in thickness) exhibits pinholes and cracks. Thus, while foils provide excellent barrier properties it is erroneous to assume these to be at the absolute zero level.

In the experimental study the foil was intentionally damaged to determine the rate of moisture penetration through these measured defects. Pouches were prepared from:

> A. 100 μ PE/adhesive/100 μ PE
> B. 100 μ PE/adh./50 μ foil/adh/100 μ PE
> C. Same as B, but with 1 pinhole/pouch
> D. Same with 2 pinholes/pouch
> E. Same with 10 pinholes/pouch

Each pinhole was 0.02″ (0.5 mm) in diameter. The presence of pinholes in the laminate could not be confirmed by visual means since black polyethylene (PE) blocked all light penetration. X-ray photos (Figure 5-2) of the finished pouches revealed the pinholes clearly—and also called attention to other minor defects which were not part of the plan.

The results of product A—the foil-free pouches—were plotted in Figure 5-1. The slope of the curve changes abruptly after the fourth week. This is not a mistake. Quite the contrary, this change in direction confirms a long acknowledged scientific axiom. For the first four weeks, the 5″ × 5″ pouches, containing 50 ml of distilled water, were kept at 70°F (22°C). The rate of moisture loss under these conditions was very low. To increase the MVTR, the pouches were placed in an oven at 140°F (60°C). In either case

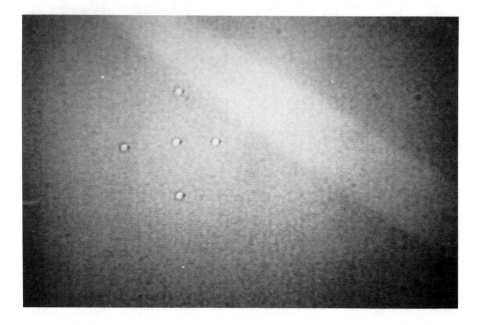

Figure 5–2.   X-ray photo of foil composite showing intentionally induced holes.

there was a liberal amount of Dryerite present to absorb humidity and thus provide a dry atmosphere surrounding the pouches.

Thermodynamics teaches us that the reaction rate "is approximately doubled or trebled for each 10°C rise in temperature" (Glasstone 1965).

During the first four weeks, moisture loss from the foil-free pouches (A) amounted to 0.9 mg/week. When the temperature was increased by 37.8°C (from 22.2° to 60.0°C), the rate of moisture loss jumped to 7.6 mg/week. This represents an 8.4-fold increase. For a temperature rise of 37.8°C one would expect a rate increase of $3.78 \times 2 = 7.56$-fold. The actual results were thus very close to the theoretical expectation.

The results for the foil pouches were very revealing, as well. The weight loss experienced by pinhole free pouches (B) was identical to the C pouches. Both types lost 110 mg in one year. This would calculate to about 0.6 mg/ 100 sq. in./24 hrs at 60°C or about 0.04 mg at 23°C. The fact that the deliberate placement of a 0.5 mm hole did not increase the weight loss, places the moisture loss through the undamaged pouches in question. Could the water migrate through the seals rather than through the walls of the pouch? In such a scenario, the loss of water through a single hole of 0.5 mm diameter may prove negligible.

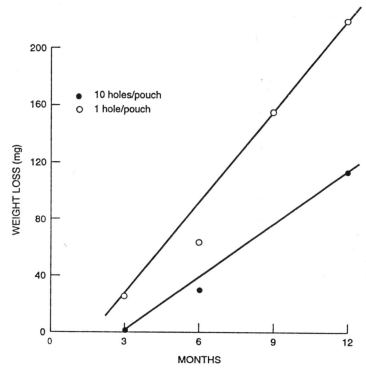

Figure 5-3. Weight loss of defective foil pouches.

pouches with 10 holes (E), 200 mg of water migrated during the year at 60°C. This amounts to 0.55 mg/24 hrs/50 sq. in.—or 1.1 mg/100 sq. in./ 24 hrs at 60°C. Extrapolating to a single hole at 23°C, yields an amount of 0.007 mg/100 sq. in./24 hrs at 23°C. This value of 7×10^{-6}g might be further reduced to 3×10^{-6}g if one subtracted the loss through the defect free foil pouch. In either case, the moisture migration is not zero—but certainly approaching same. It should, furthermore, be realized that minor defects do not compromise the efficacy of a foil package.

CASE STUDY 5-2

The packer of a liquid antacid wanted to present same in a unit dose format. The package would be similar to small cups currently available in many restaurants for the serving of cream or dressing. Since such unit dose antacid would be dispensed primarily in hospitals or other institutions, a long shelf life was deemed essential. Loss of moisture and flavor was the primary concern.

Often subtle differences in packaging materials account for major per-

Table 5-3.   Shelf Life Study, Unit Dose Antacid.

	Moisture Loss (%)	
Cup material/(lid stock)	28 days (tests)	1 year (calculated)
0.6 mm styrene/(1)	2.1	24
0.6 mm styrene/PVDC/50 μ PE/(1)	0.5	6
0.6 mm injection molded high density PE/(1)	0.03	0.4
0.6 mm foil/vinyl/(2) (supplier A)	2.1	24
0.6 mm foil/vinyl/(3) (supplier B)	0.02	0.25

Lid stock employed with above:
(1) paper/ 9 μ foil/heat seal coating
(2) nitrocellulose coating/ 65 μ foil/ heat seal
(3) paper/19 μ PE/9 μ foil/12 μ PET/ 25 μ PE

Often subtle differences in packaging materials account for major performance improvements. Several packaging materials furnished by three suppliers were tested. Results of the experiment were tabulated in Table 5-3. Since the packer desired less than 5% moisture loss during one year of storage, there were only two choices available from the range of materials tested. It is interesting to note that high-density polyethylene or a formable foil/vinyl composite performed almost equally well.

Another fact worth mentioning is the divergence of performance of foil/vinyl obtained from two reputable suppliers. It seems safe to assume that the much higher moisture loss experienced with the package from supplier A was not due to a deficiency in the cup material. It is more likely that the differences in lidding stock accounted in some fashion for the improved results obtained with product from supplier B.

ADDITIONAL READING

Anon. 1989. The best way to keep food fresh. *Consumer Reports* 54(2): 120-123.

Anon. 1987. New technology provides low cost barrier protection. *Food Engineering* 59(2): 56-57.

Brickman, C. L. 1961. Determining WVTR by new pouch method. *Package Engineering* 6(12): 311-315.

Brickman, C. L. 1962. Determining WVTR by new pouch method. *Package Engineering* 7(1): 11-17.

Glasstone, S. 1965. *Textbook of Physical Chemistry*. New York: Van Nostrand Reinhold.

Hatzidimitriv, E. et al. 1987. Odor barrier properties of multilayer packaging films at different relative humidity. *Journal of Food Science* 52(2): 472-474.

Kraus, F. 1972. Accelerated testing of shelf life parameters. *Soap, Cosmetics, Chemical Specialties* 48(11): 68-70, 86.

McCormick, R. D. 1973. Shelf life extended 5-fold by vending sandwich package. *Food Product Development* **7**(10): 15.

Sneller, J. A. 1988. Superbarriers could change the packaging rules. *Modern Plastics* **65**(5): 52–56.

Van Gieson, P. 1972. Stand-in product speeds up shelf life predictions. *Package Engineering* **17**(6): 60–63.

Q 6: What Barrier Materials Are Available?

A 6: Currently, a range of barrier materials are available. Selection of the best product for a given application requires careful consideration of many factors. Foils provide excellent barrier, but preclude product visibility. The products discussed herein offer see-through properties at somewhat reduced shelf life.

PVDC

Polyvinylidine chloride is a unique polymer. Its molecular configuration accounts for some very desirable properties. The symmetry of the molecule permits tight packing and is responsible for a high density (1.6–1.8 g/cc). Since 73% of the polymer's weight is due to chloride, the molecule is very inert. The tight packing imparts excellent barrier properties to films and coatings composed of PVDC. However, this very same tight packing is responsible for a high degree of brittleness. In order to overcome this undesirable effect, PVDC is copolymerized with PVC or acrylonitriles to improve flexibility. In the process, some of the barrier and other properties such as adhesion or sealability are sacrificed. A 0.1 mil (2.5 μ) coating of PVDC copolymer provides an oxygen transmission rate of better than 1.0 cc per 100 square inches, per 24 hours (and sometimes better than 0.5 cc). However, even multiple coats of PVDC cannot reduce the oxygen transmission rate to 0.1 cc or less. The two primary problems that plague PVDC coatings are anchorage and pinholing. The PVDC homopolymer has very poor adhesion to several substrates. Adhesion promoters are often applied prior to PVDC coating. Without this ploy one may experience "Saran lift" (the trademark "Saran®" is often erroneously employed to designate all PVDC).

When a PVDC solution or dispersion is applied to a surface, the drying operation often causes microscopic pinhole formation. These imperfections in the coating are responsible for appreciable decrease in barrier properties. Some converters attempt to overcome this deficiency by applying multiple coats of PVDC. The improvement noticed is not due to the increased thickness of PVDC, but rather due to the plugging of holes left by previous layers of coatings. One good coating could often outperform two or more poorly applied layers.

PVDC imparts many desirable properties to packaging materials. Some of the more important ones are given here.

Barrier

PVDC prevents various gases from entering into or leaving the package. The gases of primary concern are oxygen and moisture vapor. Table 6-1 provides information of the effect of PVDC coatings on several diverse substrates. It should be noted that the oxygen transmission rate of uncoated nylon at 1.3 cc/100 sq. in./24 hours or uncoated polyethylene at 129 cc/100 sq. in./24 hrs can be improved to 0.4 cc of O_2 with a thin coating of PVDC. The rate of oxygen transmission seems entirely independent of the selection of the substrate and totally controlled by the PVDC coating.

It should also be noted that while PVDC is most frequently employed to

Table 6-1. PVDC Barrier Properties.

Film	Thickness (mil)		MVTR (g/100 sq. in./ 24 hrs at 90% RH, 100°F)	OTR (cc/100 sq. in/ 24 hrs/atm at 68°F)
Polyethylene	3.0	uncoated	0.5	129
		coated	0.1	0.4
Polypropylene	0.65	uncoated	0.5	129
		coated	0.4	0.4
Polystyrene	2.0	uncoated	3.8	97
		coated	0.4	0.4
PVC	·8.0	uncoated	0.5	1.3
		coated	0.3	0.4
Polyester	0.5	uncoated	2.2	7.0
		coated	0.3	0.4
Nylon	1.5	uncoated	13.5	1.3
		coated	0.7	0.4

reduce oxygen transmission, it does in effect provide a good moisture barrier as well. It is also well to realize that moisture has no effect on the oxygen transmission rate through PVDC.

The contribution of barrier properties to the survival of packaged products cannot be overemphasized. Moisture changes and oxygen exclusion have been discussed in the answers to Q 2, Q 4, and Q 5. Flavor retention is enhanced by PVDC as well. The aroma and flavor of many foods can be protected by a barrier containing packaging material.

Grease Resistance

PVDC provides paper or other substrates with exceptional grease resistance. Packaging of oily and greasy foods such as cakes, bacon, nuts, etc. can be handled without grease penetration or discoloration of packaging material.

Heat Sealability

PVDC coatings can be heat sealed to themselves. The seals are not strong–usually just a few ounces per 1″ width. However, such seals are adequate for lightweight products such as potato chips. Many overwraps consist of single-ply printed, PVDC coated polypropylene or cellophane. Both back and ends are closed by PVDC to PVDC seals.

UV Barrier

PVDC seems to absorb UV rays in a critical region. Thus the coating protects some food items from fade (see Case Study 6–2) (Hirsch and Spiegel 1975).

Health Aspects

Dr. Cesare Maltoni has demonstrated the development of kidney cancer in rats exposed to large quantities of PVDC. The Dow Chemical Company has sponsored a more extensive study of ingestion and inhalation of controlled quantities of PVDC by rats. Thus far the study has failed to show any link between PVDC and malignancy. It should be emphasized that the data available at this writing are insufficient to permit a final judgment to be reached. However, it should be noted that the quantity of PVDC present in any package is minimal. Normally, the PVDC is trapped between layers of materials and thus not in direct contact with the packaged food. Furthermore, due to the "drying" operation associated with the PVDC application, practically all volatile PVC residues originating from the PVC/PVDC copolymer have been driven off. There is thus no likelihood of volatiles migrating into the food product.

PVDC COEXTRUSION

The Dow Chemical Company produces Saran® film, which has found only limited application in commercial packaging operations. Saran Wrap® is widely used in the home for the protective wrapping of leftovers. Its high price has encouraged competition from less costly polyethylene and PVC shrink films.

The Dow Chemical Company produces, by a patented process, a coextrusion called Saranex®. This composite film has a PVDC core, surrounded on both sides by a polyolefinic homo- or copolymer (PE, EVA, and others). This family of films exhibits good resistance to acids, alcohols, salts, glycerol, and a variety of other products. Saranex® is not recommended for the packaging of hydrocarbons and insecticides. It has found application in "bag in the box" or flexible packaging for meat, juices, tobacco, and other products.

The Dow Chemical Company has licensed others to produce PVDC coextrusions. These licensees have concentrated on heavy gauge (20 mil and up) sheeting. The cups and trays formed from these PVDC coextrusions can be found on the supermarket shelves, with aseptically packaged juices or microwavable foods (see answer to Q 8).

EVOH

Ethylene vinyl alcohol has very good barrier properties for gases, but it is very moisture sensitive and thus must be protected from moisture. While PVDC can be placed on the surface of the packaging material, EVAL® must be sandwiched between polyolefin layers to keep it dry.

Table 6-2 demonstrates the nature of EVOH resin. The oxygen transmission rate increases almost fiftyfold as the humidity rises from 65 to 100% RH. The same applies to oriented Nylon (ON). On the other hand, polyester (PET) or PVDC coated materials show no moisture dependence. Also noteworthy is the fact that Saran® HB (high barrier) can exhibit barrier properties equivalent to EVAL®. At low relative humidity, the performance of EVAL® is outstanding (Table 6-3). EVAL® has many other advantages, as well. It is relatively stable at elevated temperatures. PVDC, on the other hand, will decompose, with the evolution of hydrogen chloride, at higher heat conditions. It is thus undesirable to utilize PVDC in packaging application involving high temperature exposure.

Table 6-2.   Oxygen Transmission Rate, Barrier Properties at 20°C.*

% RH	EVAL®-F	ON	PET	ON PVDC ctd.	Saranex®	Saran® HB
65	0.02	1.92	2.9	0.35	0.45	0.09
85	0.076	5.4	2.9	0.35	0.45	0.09
100	0.97	19.0	2.9	0.35	0.45	0.09

Source: EVAL Company of America
1001 Warrensville Rd. Suite 201
Lisle, Il 60532
*Measures in cc/mil/100 sq. in./24 hours/atmosphere
ON oriented nylon
PET polyethylene terephthalate, polyester

Table 6-3.   Oxygen Transmission Rate at 0% RH and 23°C.

Film	cc/mil/100 sq. in./24 hrs/atm
EVAL®	0.013
PVDC	0.080
Barex® 210	0.800
MXDL Nylon	0.150
Oriented polyester	2.30
HDPE	150
LDPE	554
OPP	163

The data indicates that EVOH resin provides excellent barrier for aroma/flavor (Table 6-4). The material should prevent undesirable aromas from penetrating into the package and contaminating the product, but also from leaving the package and depriving the food of its characteristic flavor and aroma. Highly aromatic materials were packaged in various composite packaging materials in order to evaluate their barrier properties. Failure of the packages were measured in days and attributed to the permeation of the highly aromatic materials. The higher the number, the better the barrier (Table 6-4). Unfortunately, the evaluation lasted no more than 30 days. A number of materials survived this limited test period and cannot be ranked.

OTHER PLASTICS

PVC provides good moisture and fair oxygen barrier. This is the reason why polyvinyl chloride has been used in luncheon meat, cheese, and many other packaging applications. Unit dose cups have often been formed from PVC and their barrier has been enhanced by PVDC coatings.

Barex 210 and like plastics have been used in similar fashion.

Nylon has poor moisture but very good oxygen barrier properties. Like

Table 6-4.   Aroma/Flavor Barrier.

Composite	Gauge (mil)	Days to Leakage			
		A	B	C	D
PET/EVOH/PE	0.5/0.6/2.0	15	25	27	>30
PET/EVOH	0.5/0.6	>30	>30	30	>30
ON/EVOH	0.6/0.6	2	>30	27	30
PET/PE	0.5/2.0	2	16	5	>30
ON/PE	0.6/2.0	2	20	5	28
BOPP/PE	0.7/2.0	6	2	1	13

Source: EVAL Company of America
     1001 Warrensville Rd. Suite 201
     Lisle, IL 60532
A = vanilla.
B = peppermint.
C = heliotropin.
D = camphor.

EVAL, nylon must be kept dry to perform at its maximum potential. Thus a composite in which nylon is sandwiched between layers of polypropylene would provide good oxygen barrier, as well as the mechanical strength associated with this plastic.

INORGANICS

All of the organic plastic barrier materials suffer from several shortcomings. Most are heat sensitive, have some potential health concern associated with them, and are costly. The inorganics—titanium, aluminum and boron oxides or similar salts—have none of these shortcomings. Compared to PVDC, for example, the inorganics are much less expensive. It is not enough to calculate the cost differential on a per pound basis. There is a substantial manufacturing cost savings to be realized, as well. PVDC is normally applied in a separate coating operation. Thus PET/PVDC/adhesive/LDPE is produced in a two-step process. First the PET is coated with PVDC and then the coated PET is adhesively laminated to LDPE. The inorganic barrier material can be incorporated into the adhesive and the same composite as above can be produced in a single step—saving both material and manufacturing costs. In addition, the oxygen barrier properties of the inorganics (Table 6-5) are actually superior to the equivalent PVDC-containing composite.

CASE STUDY 6-1

A packer of frankfurters complained that the packaging material lacked PVDC as specified. His frankfurter packages were ballooning in the store.

Table 6–5.   Organic versus Inorganic Barrier.

Composite	Oxygen Transmission (cc/mil/100 sq. in./24 hrs/atm)	
	PVDC Coated	Inorganic Adhesive
Polyester/PE	0.76	0.30
Nylon/Surlyn	1.05	0.60

Source: Hirsch, A. and Lechleiter, R. 1977. PVDC replacement. Unpublished report, Standard Packaging Corp., Clifton, N.J.

This latter fact was indeed verified. However, tests on the packaging material involved confirmed the presence of PVDC.

It was postulated that the ballooning was due to bacterial activity within the meat product, generating gas. The gas was trapped within the barrier package. Elimination of the PVDC barrier permitted the gas to escape through the package wall. The gas generation continued, but was no longer observable.

This was an expedient solution to a problem—without resolving the serious causes thereof.

CASE STUDY 6-2

In the wake of restrictions placed on the use of PVC (due to retention of residual monomer), there was concern that the FDA might ban the use of PVDC as well. With the cooperation of L. A. Frey, a meat packer, a study was undertaken to evaluate a potential inorganic replacement for PVDC.

Bacon and bologna were vacuum packed in Nylon/Surlyn and polyester/Surlyn composites containing the inorganic barrier material incorporated into the adhesive. The processed meat was packaged on the standard equipment under normal operating conditions. No problem was encountered.

Packages were placed in refrigerator at 38°F for 45 days. No packaging failure, delamination, bond falloff, or other signs of package deterioration was noticed.

Product packaged in experimental film was exposed to UV light for 45 days. Absolutely no color change was effected by this prolonged exposure.

ADDITIONAL READING

Anon. 1984. Barrier materials reshape packaging. *Food Engineering* **56**(9): 96.
Anon. 1988. Diverse technologies are emerging for coextruded barrier packaging. *Modern Plastics* **65**(7): 30.

Anon. 1988. PET/EVAL bottle has 9 month shelf life. *Plastics Technology* **34**(2): 101.

De Lassus, P. T. 1988. Barrier expectations for polymer combinations. *TAPPI Journal* **71**(3): 216–219.

Harada, M. 1988. A new barrier resin for food packaging. *Plastic Engineering* **44**(1): 27–29.

Hirsch, A. and Spiegel, F. X. 1975. PVDC prevents fade of processed meats. *TAPPI PVDC/Adhesives Laminating Seminar.*

Lantos, P. R. 1985. Polymer blends and alloys will upgrade food packaging. *Packing* **30**(5): 59–62.

Manne, S. 1976. Barrier coated plastics improve product protection of thermoforms. *Packaging Digest* **13**(1): 49–51.

Nouni, Cl. 1974. Problems with PVC coating of polyolefin films. *Verpackungs Rundschau* **74**(17): 10–13.

Perdikoulias, J. and Wybenga, W. 1989. Designing barrier film structures. *TAPPI Journal* **72**(11): 107–112.

Rice. J. 1987. Tough polyester blends offer user economics plus oxygen, moisture, flavor and aroma barriers. *Food Processing* **48**(2): 152.

Russell, M. J. 1989. Barrier plastics. *Food Engineering* **61**(5): 89.

Sacharow, S. 1972. PVDC film in packaging. *Flexography* **17**(9): 20–42.

Schaper, E. B. 1989. EVAL® copolymer resins for better solvent, aroma and flavor barriers. *TAPPI Journal* **72**(10): 127–131.

Q 7: Is Foil Necessary?

A 7: In some packaging applications foil is the one and only choice available for a high-barrier flexible packaging material. Foil entails two important disadvantages which compel a thorough review of the real necessity for this type of package. Foil packages are first of all very expensive and furthermore preclude visibility of product. These two shortcomings must be considered prior to the selection of this type of packaging material.

The transmission rate of both oxygen and moisture can be drastically reduced with the aid of foil. It is erroneous to assume that the use of foil in the packaging material will reduce the oxygen and moisture transmission rate to the ideal zero.

Both, because of cost and flexibility considerations, one employs the thinnest foil material suitable for the desired application. Very thin gauges such as 0.00030" or 0.00035" of foil (7.5–8.75 μ) have minute pinholes and are thus not the absolute barrier that one imagines them to be. As the foil gauge increases the probability of pinholing diminishes. Some packaging technologists prefer to utilize two plies of 30 ga. rather than a single ply of 70 ga. aluminum foil. They argue that the probability of aligning two pinholes in a lamination of two plies of a thin gauge foil is much slimmer than that of finding a single pinhole in a 70 ga. foil product.

Aluminum foil is available in a wide range of alloys and tempers. The most commonly chosen alloys are 1145 and 1100 foils, which are generally utilized in flexible packaging, and the 3003 foil, which is used for semi-rigid containers. These numbers correspond to the alloy compositions. The tempers range from fully annealed or soft foil, designated 0 temper, to hard foil, designated H 19. Annealed foil is normally selected for the most flexible packaging applications since it possesses the highest degree of formability. Intermediate tempers, such as H 25 and H 27 are employed in formed contour applications.

One problem encountered in the processing of foils is caused by the presence of lubricating oils. Most of these residual rolling lubricants are removed from the foil surface in the annealing process. Consequently it is usually possible to obtain good adhesion onto the foil surface. However, the presence of oil residues can play havoc with the adhesion of inks and/or coatings onto a contaminated foil surface. This happens more frequently with harder foils.

The need for foil will be determined by the shelf life expectancy of the product and its maximum permissible moisture loss or gain, as well as oxygen susceptibility. Very often, however, data on maximum permissible moisture loss or gain are not available. Table 7–1 compares moisture vapor transmission rates (MVTR) of a variety of packaging materials on a 24-hour basis. To attain the actual moisture gain or loss for any given package during its desired life cycle, one must multiply the results shown in Table 7–1 by the number of square inches in the total package and furthermore by the number of days of shelf life expectancy.

Foil packaging is quite prevalent in the pharmaceutical industry, where maximum protection and prolonged shelf life are of prime importance. In the food industry, foil packaging is reserved for soup mixes or other dry powdered products (such as dehydrated milk) which are very sensitive to moisture and are stored at room temperature on the open shelf.

Foil packaging has found a new application in the "flexible can" where the food product is actually sterilized in the pack and stored at room temperature. More about this application is an answer to a subsequent question (Q 19).

Table 7–1. Moisture Vapor Transmission through Select Packaging Materials.

Composite	MVTR (gm/100 sq. in./ 24 hrs at 23°C)
195 cello/0.001″ PEX/35 ga. foil/0.002″ PE	0.0023
195 MSBO cello/35 ga. foil/HSC	0.0095
0.0005″ OPP/250 K cello/0.002″ PE	0.12
195 K cello/0.002″ PE	0.26
0.0005″ PET/0.002″ EVA	1.14
0.001″ Nylon/0.002″ EVA	2.2

METALLIZED FILMS

In recent times, metallized packages have gained some popularity, especially in Europe. To better understand the possible application for these packaging materials, it may be best to review the metallizing process at first.

The Metallizing Process

The process of metallizing of plastic films produces a thin uniform layer of the coating metal on a continuously moving web of film. This is done by heating the metal in a high vacuum and allowing the resulting metal vapor to condense on the moving web. Metallizing is primarily a semi-continuous process, i.e., just one reel of film at a time is contained in the vacuum chamber. Continuous metallizing, in which reels of film are treated successively without interruption, can be processed—however, it requires a more elaborate system of vacuum chambers and pumps, which adds to the capital cost of the equipment.

Aluminum is the metal typically employed, although copper, silver, and gold can be deposited by the same process. Zinc can also be coated onto plastic films by a slightly different process; a thin silver layer must be deposited first to nucleate the zinc.

The critical part of the metallizer is the vacuum chamber, which contains:

1. Equipment for unwinding and rewinding the film. This includes devices that control web tension, flatness, and speed.
2. A heated source emitting metal vapor. Usually the source is an electrical resistance-heated "boat" containing a small quantity of the molten metal. These boats are consumable items and contribute significantly to the running costs of the metallizer. Each boat has a mechanism for feeding aluminum wire at a rate which just replenishes the losses caused by evaporation. Other methods of evaporating the metal include the use of induction-heated crucibles or electron beam evaporation. Cathode sputtering has also been used for metal deposition.
3. A cooled drum over which the film passes while metal is deposited thereon.

Figure 7–1 depicts schematically a section through a typical vacuum metallizing chamber and illustrates the principle of operation.

The vacuum is attained by way of a large diffusion pump backed by high-speed rotary pump systems and mechanical boosters. Since the time required to achieve a working vacuum (less than 10^{-4} torr) has a direct bearing on the operating cycle and hence on the economics of the process, this

Figure 7-1. Vacuum metallizer. (*Courtesy John Dusenbery Co., Inc.*)

pumpdown must be as short as possible. It usually lasts 10 to 15 minutes. The low outgassing characteristics of polyester film are of special significance in the pumpdown phase. For polyester film a single-chamber vacuum system is normally adequate. However, multiple-chamber systems, which do not need large, expensive diffusion pumps, can be employed to reduce pumpdown time. Certain applications call for metallization of both sides of the film—and equipment is available which accomplishes this feat in a single operation.

The size of the vacuum chamber selection is dictated by the maximum width and diameter of the film reel to be metallized. Metallizing chambers accepting reels of up to 79 inches (2 meters) in width are available, but equipment for widths of up to 60 inches (1.5 meters) is more common. The accessibility of the vacuum chamber is of primary concern, especially if large reels are to be loaded and unloaded with mechanical assistance.

The thickness (typically 300 Å) of metal deposited on the web is determined by a number of factors that include the power input to the heaters, the pressure in the vacuum chamber, and the web speed. In practice, adjustment of web speed is the more usual method of varying the overall thickness of the deposited metal. Variations in the coating across the film web can be corrected by adjustment of the power input to the individual heaters. Thickness can be monitored visually, by photoelectric devices, or by measurement of electrical resistivity. Typical values of the latter range from 1 to 10 ohms/square.

Table 7-2. Light and Oxygen Barrier Properties of 0.0005″ (12 μ) Polyester.

Ohms/sq.	Light Transmission		O₂ Measure (cc/100 SI/24 hrs)
	% Visual	% UV	
Unmetallized	93.9	91.2	7.0
55	46.8	48.9	1.5
15	35.8	46.77	0.37
10	22.8	42.66	0.294
4.5	1.58	4.17	0.105
3	0.603	3.47	0.09
1	0.0005	0.0005	0.02
Two-side			
5.0/side	0.034	0.155	0.07

Performance Characteristics

It is claimed that metallized polyester film has excellent light, oxygen, and moisture barrier properties. The performance data is assembled in Table 7-2. It relates ohms/square measurements to a deposition parameter. The lower the ohm reading, the lower the resistance to electric current flow, and therefore the better the coating. This assumption, however, is subject to serious question and review. It was found that visible holes could be punctured into a metallized film without influencing its ohm performance. Thus the correlation established in Table 7-2 needs to be reexamined.

It is admitted that the best metallized film does not offer barrier properties equivalent to those of a foil composite. However, many products packaged in foil nowadays do not really require an oxygen transmission rate of less than 0.01 cc/24 hours. Since the metallized polyester composite is less costly and machines better due to its flexibility it should find application where the extreme barrier of foil is nonessential.

The claim as demonstrated in Table 7-3, that metallized polyester is actu-

Table 7-3. Oxygen Barrier before and after Flex Tests.*

Composite	Flat	Flexed 100 Cycles
0.0005″ metallized PET/0.002″ PE	0.09	0.28
250 K cello/0.0005″ foil/0.002″ PE	0.00	725.00
250 K cello/PX/foil/0.002″ PE	0.00	240.00
Paper/PX/foil/0.002″ PE	0.00	235.00

*100 times on Gelbo Flex Tester.
PX extruded low density polyethylene.

ally superior to a foil composite, is accepted only with reservation. It is argued that in actual use, packages are flexed. In a simulated flex test the metallized packaging material retained far better barrier properties than the flex-cracking-prone foil composites.

Applications

The primary target for metallized packaging materials is consumer impulse buying. The high sheen of metallized polyester, combined with good graphics, will result in an appealing package that will attract the average shopper. This marketing approach does work (as proven by several products on the market). However, this packaging advantage diminishes as more and more competitors adopt the same packaging mode. The uniqueness of a metallized package is important in its sales appeal.

In Europe as well as in Canada, strip metallization has been utilized. Rather than have the entire package or one face fully metallized, one observes just portions of the package surface covered with aluminum. This type of pack looks as if it had an aluminum label affixed to one of its faces. It provides the package with product visibility while at the same time making advantageous use of the mirrorlike appeal of a metallized composite.

FORMABLE FOIL

In theory all soft foil is formable. Practically, however, cold forming of thin gauge foil is difficult if not impossible. Recently, foil/plastic combinations have been introduced to permit the forming of such composites on standard flatbed machines into good-sized trays. This packaging has gained limited acceptance but is viewed with renewed interest by food packers.

Another important outlet for formable foil/plastic composites is in the drug market. Suppositories are currently packaged in this manner and other applications are likely.

The readers attention is called to Q 4, Q 5, and Q 6, dealing with various aspects of barrier packaging, including foil utilization.

ADDITIONAL READING

Adams, F. 1976. Taking the mystery out of metallizing. *Research and Development* **76**(2): 44.

Anon. 1988. New approaches for foil lids. *Packaging* **33**(3): 85.

Anon. 1976. Flexibles: high barrier for fresher foods. *Modern Packaging* **49**(7): 21–24.

Anderson, A. N. 1988. Typical causes of pinholes in pack rolled foil. *Lubricating Engineering* **44**(7): 622–625.

Ashton, R. 1989. Packaging growth will slow in 1990. *Packaging* **34**(12): 50–51.

Bear, C. A. 1974. Vacuum metallizing—materials, process, economics. *Research and Development* **74**(2): 44.

Keski-Kuha, R. A. M. 1989. X-ray metal film filters at cryorgenic temperatures. *Applied Optics* **28**(7): 2965–2968.

Newton, J. R. 1977. Polyester as a metallized substrate. *Modern Packaging* **50**(1): 38–40.

Radtke, R. 1976. Comparative research of potato chips during storage in aluminum/plastics and cellulose/plastic film combinations. *Verpackungs Rundschau* **27**(10): 79–82.

Reichert, J. E. 1975. Latest knowledge of producing canned salt potatoes in glass, can and film or foil containers. *Verpackungs Rundschau* **26**(9): 1130–1141.

Uetzmann, P. 1989. Aluminum foil. *Packaging* **34**(5): 54–55.

Wagner, J. N. 1984. Metallized film could be a foil substitute. *Food Engineering* **56**(6): 66–67.

Q 8: Why Use Rigid Plastics?

A 8: Plastics have found wide acceptance as packaging materials. These modern replacements for wrapping paper fulfill all the packaging requirements: contain, protect, and display—and all that at a relatively low cost. A further innovation of packaging utilizes "rigids" to add quality and a variety of new dimensions to the packaging function. Rigids can be shaped to the contour of the product and thus provide improved visibility.

The rigid blister affords often better protection for the product thus packaged. Such increased protection may be furnished in several ways. The rigid can often provide better oxygen and moisture barrier. It can also reduce package failures associated with shipping and handling. The glasslike appearance of rigids affords improved display, as well.

Rigids, nevertheless, are not a cureall. These plastics must be utilized judiciously. Rigids are normally more expensive than all-flexibles, and this increased cost must be justified. The extra packaging cost must be dictated by competitive pressure, by a desire to convey a higher quality image, by the need for special protection, or some other valid reason. Occasionally one encounters the unjustified indiscriminate use of rigids.

The term "rigids" requires further definition. There are many such plastic materials available. Table 8–1 covers a wide property and price range of rigids.

Polystyrene is one of the least expensive materials of this type. Because of its poor impact properties, it has limited utility as a protective packaging material. In spite of its durability limitations, there are many uses for various grades of polystyrene. However, in most instances this material is utilized as a display cover, involving little or no shipping endurance requirement.

PVC (polyvinyl chloride) has been the most widely used rigid material

Table 8-1.   Properties of Rigid Plastics.

Properties	Styrene	PVC	Barex	PET	Kodar
Gauge (mil)	10	10	10	10	10
Yield (SI/lb)	2650	2100	2400	2100	2200
Density (g/cc)	1.05	1.35	1.15	1.33	1.23
Tensile (psi) MD/TD	12000/12000	6600/6700	8500/8000	11200/11100	7000/7600
Elongation (%) MD/TD	3/3	220/200	90/90	430/430	200/215
WVTR (g/100 SI/ 24 hrs/mil)	1.0	0.3	0.6	0.42	0.49
OTR (cc/mil/ 100 SI/24 hrs)	100	1.8	0.08	1.15	2.5

SI square inches.
MD machine direction.
TD transverse direction.

in the food and drug industry. Concern for possible cancer causing agent associated with residual VCM (vinyl chloride monomer) has prompted a search for alternatives. Improved PVC has retained a large portion of the traditional business. However, alternatives must be considered, since they will continue to play an important role in packaging.

PET (polyester), PAN (polyacrylonitrile), XT (polyacrylate), and others are all actively competing for this market. PVC has the advantage of having an established history of usage and success. Furthermore, most packaging machinery in current operation has been adjusted to run PVC sheeting. Users of the various alternatives have encountered difficulty in processing these materials on their packaging equipment. Those with a determination to eliminate PVC from their operation have learned to handle these materials. Others have made a pro forma attempt and have switched back to PVC.

The cutting difficulties encountered with PET at one time eliminated it from serious consideration in the market place. Co-PET (Kodar®) has found wider acceptance but suffers from an uncertain supply situation.

Polyacrylonitrile, which at one point seemed the ideal PVC replacement, has come under FDA scrutiny as well. Its use in plastic bottles has been banned. A major producer of this resin has withdrawn its version of this product from the market. This leaves a diminished commercial interest to battle FDA, PVC, and all other rigids.

More costly rigid plastics are available for specialized applications. High-temperature-resistant plastics are an example of this class. Polycarbonates, polysulfones, HDPE (high density polyethylene), and polymethylpentene all exhibit different degrees of high temperature resistance together with a variety of other desirable properties. It should be remembered that rigids, like flexibles, may be furnished as composites as well. Thus one can combine one or more rigid layers with one or more flexible plastic layers to create a composite affording a unique array of properties. One should further consider that most rigids are thermoformed. Properties are often determined for the rigid "as is," and may be seriously altered upon forming. It may be desirable to reevaluate the performance characteristics of the thermoformed package.

PORTION PACK

Many items served in the restaurant or sold in the supermarket are prepackaged in a measured quantity to be served at one meal only. Jams and jellies, for example, are served in small containers (usually about 1/4 oz.). The package is attractive as well as hygienic. Unopened containers can be returned to the storeroom for future use. The same applies to butter, cream, salad dressing, and a long list of other items. The package cost has been more than justified by waste reduction and increased customer satisfaction.

UNIT DOSE

The experience of the food service industry has been applied with some success to the dispensing of medication. Hospitals that suffer from shortages in trained personnel have found that delivery of premeasured and prepackaged medication to the patient is cheaper and involves less risk of error. PVDC-coated polystyrene cups are often adequate for this task.

MICROWAVABLE

Interest in microwavable packages has been sparked by the general availability of these heating/cooking devices in the average home. It is estimated that the majority of all American homes are equipped with a microwave oven. This fact presents new challenges to the food packager. (For a detailed discussion of this topic see Q 31.) With regard to rigid packaging, there are several unique problems. The package may have to withstand freezer storage and the heat generated in the microwave oven. Even if the plastic remains relatively unaffected by microwave energy, the contents—such as soup—could reach boiling and in turn raise the package temperature. The plastic container must retain its shape and rigidity at the elevated temperature, especially if it is to function not just as a storage vessel but also as a serving tray.

OVENABLE

The problem discussed above is magnified in an ordinary convection oven. The exposure in the microwave oven is limited in duration to just a few minutes. The surrounding atmosphere is at room temperature and the contents are normally limited to below 200°F (90°C). In the standard kitchen oven, the air temperature may range from 250° to 450°F (120° to 230°C) and the exposure may last up to 90 minutes (rarely longer and very often the heating will last less than 45 minutes). The thermoformed package serves as a cooking or baking tray and must retain its shape and appearance. Details of heat-resistant packaging are discussed in response to Q 31.

BARRIER PACKAGES

Rigid composites are available with excellent barrier properties (see the answer to Q 6). These materials have made it possible to aseptically package items such as orange juice. More about this subject matter under Q 11.

CASE STUDY 8-1

A packer of sliced ham complained of excessive flex cracking of his PVC packages. Investigations revealed that the product was undersized.

It was pointed out to the meat packer that in a semi-rigid vacuum package, the product must fill the formed tray portion of the pack fully. If the product is undersized the package walls will collapse around the voids. This will create a very unfavorable visual effect, detracting from the primary objective of rigid plastics—improved display. Furthermore, the crumpled appearance leads eventually to cracks in the plastic which destroy its protective function as well.

Oversized product may prove equally troublesome in a vacuum package. Loading the preformed PVC tray may slow down the operation or may generate a high percentage of rejects.

The packer was eventually convinced that if product size control was unattainable, then the best alternative was the changeover to an all flexible package.

CASE STUDY 8–2

During the mid-1970s a link between VCM and liver cancer was discovered. Government pressure and adverse publicity induced many buyers of PVC packaging materials to seek alternative plastics. This is the actual odyssey of one major meat packer.

This meat packer had several packaging machines at four locations throughout the USA, vacuum packing processed meat in semi-rigid PVC. Following an announcement by Oscar Mayer that this leading meat processor was dropping PVC in order to safeguard the consumer, our man decided to follow suit. PET, PAN, and XT were tested. On this limited test the PET performed best. However, in the next few weeks, as his operation switched 100% to PET, the production problems multiplied. First, there were cutting problems. The ordinary knives did not suffice to separate linked packages. After a while this perplexing situation was resolved by special knives. Also, die trimmers seemed to wear more rapidly. PVC lines permitted die trimmer utilization for up to 1.5 million strokes. On PET material, the die appeared to dull after 300,000 strokes. The changes in dies thus necessitated were costly and time consuming. Microscopic examination of the die revealed no wear, no rounding of corners. Many studies were conducted by resin and film suppliers, but the puzzle of the die trimmer failure was never resolved. Eventually, one of the prime PET suppliers withdrew the product from the market.

Our meat packer blamed his die trimmer problems on the switch to PET. For a year he struggled along with resharpening dies at a fierce rate. But eventually he gave up and searched for another replacement. Parenthetically, it should be noted that this same meat packer had severe die trimmer problems for years, even when PVC was in use. Some of his dies lasted no more than 50,000 strokes and most had to be replaced at under a quarter million cuts. However, we all have memories of convenience—we select what we like to remember.

Our meat packer tested Barex 210, an acrylonitrile copolymer. Product exhibited some cracking due to cutting and possibly lower impact resistance. But it did trim well and seemed to perform satisfactorily in every other respect. Just as a switch to Barex was about to be decided, the FDA banned the use of Cycopac (another acrylonitrile resin) for beverage bottles. While this prohibition had no direct bearing on meat packaging, it was nevertheless considered prudent to avoid the use of acrylonitrile under these circumstances.

CoPET (Kodar) was tested next and performed generally well. The die trimmer problem was less pronounced, although not entirely eliminated.

The economics of CoPET was not favorable when compared to PVC. After some time, the packer of this case study decided to return to PVC and continue its use until FDA would enforce a change.

This case study is not typical of what happened throughout the industry. It is just an example of one specific case.

ADDITIONAL READING

Anon. 1988. Ready meals convenience demand dual-ovenable performance. *Packaging* (UK) **59**(4): 24–26.

Anon. 1988. New consumer breed spurs rigid plastics. *Canadian Packaging* **41**(2): 16–20.

Anon. 1971. ABS scores in packaging. *Plastic World* **29**(5): 50–52.

Anon. 1973. Flexible packaging looks at rigid markets. *Industry Week* **177**(5): 42–47.

Anon. 1987. Dual-ovenable trays. *Foods* **156**(1): 122.

Bruck, C. G. et al. 1976. Migration of vinyl chloride from PVC packages. *Fette-Seifen-Anstrichmittel* **78**(8): 334–337.

Day, M. R. and Newton, J. R. 1976. Polyester bid to increase packaging applicationa. *Modern Packaging* **49**(2): 28–33.

Erickson, G. 1990. New trends make food a challenging game. *Packaging* **35**(1): 44–48.

Hickox, B. 1977. FDA judge due to issue first key ACN decision. *Food and Drug Packaging* **37**(3): 1–19.

Lane, W. A. 1977. How to choose the most efficient thermoform packaging system. *Package Development* **2**(5): 20–25.

Merlin, J. W. 1986. Plastic packages. *Consumer Research Magazine* **69**(7): 38.

Papaspyrides, C. D. 1986. Some aspects of plasticizer migration from PVC sheets. *Journal of Applied Polymer Science.* **32**(7): 6025–6032.

Serchuck, A. 1977. Polycarbonate has its time come? *Modern Packaging* **50**(10): 39–41.

Stouffer, L. 1988. Perspective: new worry for PVC: Its food packaging status. *Packaging Digest* **25**(5): 12.

Q 9: What Types of Package Failure Are Encountered?

A 9: Package failure for whatever reason is undesirable. Yet mechanical, material, or personnel imperfection will be encountered. These are transmitted to the package in one form or another. Package failure can be classified into several groups:

IN-LINE FAILURE

It is often quite obvious that the package as made is unsuitable for shipment. The product is not contained properly—it spills from the imperfect package. Causes may be:

1. Mechanical. Check filling equipment—it could be defective.
2. Package feed could be out of synchronization.
3. Closure. Sealing equipment could be defective or misaligned.
4. Packaging material may not be suitable for the machine or the product to be packaged. On the other hand, the material could be defective.
5. Personnel. Often package failure is a "people" problem. Machines are not properly maintained, the operator is sloppy, the inspectors are negligent, etc.

Frequently, in-line failure is due to a combination of the above factors. Correction of one single deficiency may alleviate the package failure, but normally not for long. The basic attitude toward running an efficient packaging operation determines its ultimate success.

DELAYED FAILURE

Package may come off the equipment in what seems to be perfect condition. Within a matter of hours or days, prior to shipping, the package fails. In most instances one may assume that the defect was present at packaging time but went undetected. The search for causes should be similar to those discussed above.

SHIPPING FAILURE

Some limited failure due to shipping is practically unavoidable. Even cans do split occasionally. If failure exceeds 1%, then causes should be investigated. Very often the primary reason for failure is inappropriate handling during shipment. Examination of shipping containers should produce evidence of such abusive handling.

Not all shipping failures are due to excessive impact. Shipping containers may have been exposed to the elements (rain, freezing, or high temperatures) contrary to shipper's instructions. The appearance of the external container is not always indicative of such abuse. Yet widespread failure of packages (especially those located near the periphery of the shipping container) would lead one to presume such negligence.

Shipping failure may be due to an inherent package defect. Material selected may be inadequate for normal abuse associated with the packaged product. Thus a packaging material which performs well for bologna may fail to contain hard salami. Failures could be induced by the abrasion or cutting action of the product. The material selected might be quite suitable for limited-radius distribution and yet develop serious problems when shipped over longer distances.

RETAIL DISPLAY

Most packages arrive at the retail store in perfect condition. Defective packages are easily segregated prior to store display. Yet many of these packages that have survived the abusive distribution cycle fail in the supermarket or in the hands of the consumer. To shift the blame for these failures to the consumer for mishandling the package seems self defeating. It is no doubt true that the consumer, with or without intention, squeezes and punches, drops and tosses and in many other ways places unusual demands on the package. However, our present retailing system with open shelf display requires such exposure and the packer has to accept it and compensate for it in improved packaging.

QUALITY ASSURANCE

The primary concern should deal with package failure prevention. However, since cost considerations do not permit a zero failure level, one must also consider failure detection. A good detection system can avoid getting defective packages into the distribution system.

To avoid failures of any sort one must pay attention to many details, including:

Packaging Machinery Selection

Whenever possible, choose equipment that has a proven record of performance, requires a minimum of maintenance, etc.

Packaging Materials Selection

Choose material that will be compatible with the product to be packaged and will perform flawlessly on the equipment selected. Far too often the packaging machinery and material are chosen by different departments—neither considering the need for interaction.

Maintenance

Most package failure is attributable to poor maintenance. Equipment must be lubricated and cleaned, parts must be replaced, belts must be tightened, etc. If these chores are not performed routinely then failures are inevitable. Proper maintenance attitudes are contagious and will inspire operators and other packaging personnel to exercise care.

Packoff and Shipping

Thought must be given to all the steps following the packaging operation. The perfect package is often injured through thoughtless posthandling. One cannot drop a package of potato chips from one level of the building to the next lower level, a distance of eight or more feet, and wonder at the high failure rate. The same package carefully conveyed the identical distance will survive unscathed.

QUALITY CONTROL (QC)

Failure prevention and detection requires adequate quality control at all levels. The quality of incoming raw materials must be tested. This normally involves tests designed to simulate stress and strain endurance as well as packaging operation performance. QC must check packages as they come off the packaging equipment and at various intermediate operations on the way to the market. It is not adequate to verify that the package off the machine is passable. Good quality control procedures will include the retention of samples for aging studies and adequate recordkeeping to assure traceability of packages and product. It must be remembered, however, that most quality control tests are imitations of a real event, at best. A box of packages riding on a shake table may absorb punishment similar to the carton transported on a truck. It is, however, not the identical experience. Failure may develop sooner or later or in an entirely different manner. Finally, it must be remembered that most quality control tests are destructive in nature. A package's endurance is tested by subjecting it to forces beyond its survival expectancy. In this manner a failure is induced at or beyond a predetermined performance level. This fashion of test must of necessity be restricted since 100% testing would result in total destruction of all packages. One must rely on statistical analysis—and zero failure level is virtually unattainable.

In short, thought and care can drastically reduce, if not totally eliminate, package failure. Since some degree of failure is destined to remain, one must provide for means of early detection in order to take corrective action and thus eliminate recurrence.

FAILURE DETECTION

Ideally, each package should be examined and its integrity verified. Such test(s) would of necessity have to rely on nondestructive procedure(s). The test would also have to be high speed in order to keep pace with the large number of packages generated by modern packaging equipment.

Some failure testing equipment is available right now. It is, however, at best an indication of things to come. Examples of such systems are described below. They are selections of what is on the market at this time. But this list is certainly not all inclusive. New methods and instruments are likely to surface as this manuscript goes to press.

Sniff Procedure

The package is filled with a minimal quantity of tagged gas. The term "tagged" denotes an organic substance which has been synthesized with an abnormal isotope such as Carbon 14 (instead of carbon 12) or nitrogen 15 (in place of nitrogen 14). The gas injected into the package is inert and is present in minute amount. The finished package is passed into a "detection chamber." The chamber is closed and vacuum is applied. The emission air is tested by a sniffing device for the presence of the tagged gas. Its presence is taken as evidence of package failure. The system unfortunately suffers from a multitude of shortcomings. It is applicable only to packaging machinery which can inject a measured quantity of gas. Most packaging equipment in current use could not meet this requirement. The system would also be inapplicable to vacuum packaging, where even minute amounts of gas would detract from the desired appearance. The evacuation and sniffing cycle require considerable time and thus slow down the entire packing operation. Even a slow packaging machine delivers 30 packages per minute. This would necessitate multiple sniffing lines. A high-speed line of about 1,000 packages per minute would be forced to resort to statistical sampling and thus defeat the concept of 100% inspection.

Lastly, the sniffing device is subject to rapid fatigue. The testing equipment is thus not very reliable when employed on a continuous high-speed line.

Chemical Color Changes

It is claimed that a color indicator is available to detect package failure. The system depends on the injection of a controlled amount of carbon dioxide into the package. The color of a dot or stripe of indicator ink on or in the package depends on the CO_2 concentration. Thus, if the package permits some of the carbon dioxide to escape, a color change will be noticed. The system has some merit. If it were to function properly it would provide a permanent detector, which would protect the consumer from purchasing a defective package. In this respect it would be superior to the sniff method. The latter assures package integrity only to the moment of sniffing. But the color indicator performs its function until the consumer opens the package. However, the chemistry of this system is by no means foolproof. The speed of the reaction has yet to be established. The proposed method requires gas injection—a shortcoming that has been discussed above.

Seal Imperfections

A large proportion of flexible package failures are due to seal imperfections. The Army Natick Laboratories has devoted a great deal of attention to this problem. Others have patented the use of color intensity to disclose seal imperfections. No simple, foolproof method of rapidly identifying seal defects is in commercial use.

CASE STUDY 9-1

At a national meat packer, the bacon packages coming off one vacuum packaging line developed an unusually high leaker rate. These package failures are easily detected since good vacuum packages have very tight cling, while "leakers" indicating loss of vacuum, are loose. Examination of packages revealed tears, punctures, or rips in a random pattern. One could safely exclude machinery related failure. An equipment malfunction would normally generate a deficiency in a repetitive pattern.

The material was tested and found to meet performance specifications. Since similar materials performed adequately at many other packaging installations, it was decided to seek the failure cause elsewhere. Close observation of the entire operation soon revealed the problem. A young lady with a very recently acquired engagement ring handled a portion of the finished packages coming off the end of the line. The exposed diamond punctured or tore the package inadvertently as it brushed against its surface. By turning the diamond to face toward the palm, package failure was reduced to normal proportions.

CASE STUDY 9-2

A small meat packer, vacuum packing brine-pickled briskets, reported a very high shipping failure rate. Briskets of about two to seven pounds were vacuum packed in a flexible packaging material. Six of these individual packs were placed in a carton. This procedure was similar to that followed by several other packers of briskets. None reported shipping problems. An on the spot observation revealed that:

a. The shipping department did not follow instructions of placing six packages in a carton. The number of packages varied depending on how many could be "jammed" into the carton. This stuffing operation itself could be the cause of the high failure rate.

b. Cartons were neither glued nor tape sealed but were stapled. Since the packages more than filled the carton, the staples had to puncture at least two of the top packages. Thus there was an assured minimum failure rate of 25% (two out of eight) built into this procedure.

Elimination of the above erroneous packaging method did indeed reduce the shipping failure drastically.

ADDITIONAL READING

Anon. 1988. Visual inspection system focuses on total quality control. *Food Engineering* **60**(1): 91–92.

Anon. 1976. Consumers don't complain about defective packages, they stop buying. *Food and Drug Packaging* **34**(4): 1–16.

Bartnick, A. 1988. Implementing SPC: Koch Label's story. *Converting* **6**(4): 50–60.

Burgess, G. J. 1988. Product fragility and damage boundary theory. *Packaging Technology and Science* **1**(1): 5–10.

Gonnella, C. J. 1973. Achieving defect free packaging through prevention and correction. *Package Development* 3(4): 8–17.

Kehres, L. 1988. The quest for quality control. *In-plant Reproduction and Electronic Publishing* 38(4): 42.

Kelsey, R. J. 1976. Unblinking electronic eyes, sensors and brains displace human inspectors in packaging control. *Food and Drug Packaging* 34(11): 4–26.

Kelsey, R. J. 1988. Robots in packaging. *Food and Drug Packaging.* 52(3): 45–47.

Larson, M. 1988. On-line inspection offers simpler choices. *Packaging* 33(1): 68–71.

Mansur, R. T. 1988. Nondestructive high speed leak testing for retort pouches. *Activities Report of the R & D Associates* 40(1): 48–49.

Miller, A. 1986. Giving packaging a bad name. *Newsweek* 108(7): 45.

Skrzycki, C. 1986. Tampering with buyer confidence. *U.S. News and World Report* 100(3): 46–47.

Trombly, J. E. 1988. How to specify a machine vision system. *Journal of Packaging Technology* 2(2): 50–51.

Q 10: What are the Prime Factors Going into the Selection of a Flexible Form-Fill-Seal Package?

A 10: Preformed pouches are widely used in the packaging of food items. As the volume of product increases, manufacturers turn to form-fill-seal (ffs) equipment to meet the demands of increased productivity. If flat items such as slices of luncheon meat are involved, they can be packaged on machines that form flat pouches around the product. However, high-profile products are not suited for such package formation. They must be placed in a tray or a formed blister. Many such items are packaged in rigid trays (10 mils +), either preformed or made in line. The cost of such a package is very high and cannot be justified in many instances. The use of a flexible forming web reduces material cost and speeds production.

One is confronted with a multitude of seemingly complicated machinery in packaging. Yet closer examination of diverse equipment reveals certain basic concepts shared by all. The complexity arises from the refinement of these fundamental designs. Packaging machines normally perform three functions:

1. Make the container (Form).
2. Place the product into the container (Fill).
3. Close the container (Seal).

Many other operations may be associated with the packaging operation and thus become part of the automated packaging line. For example, slicing, counting, or preweighing are all preparatory to packaging. Nevertheless, many machines will incorporate these operations into their total packaging system. The same may apply to code dating, pricing, labeling, etc. Evacuation or gas filling should probably be considered as a modification in the closure procedure. A very brief description of each of the three basic packaging operations is offered here.

FORM

A container is prepared from either one or two webs. The most common types are either pouches or formed shapes.

Pouches

On a horizontal machine, a single web can be folded into a U shape, with its open end up. It can then be sealed into individual compartments, yielding attached envelopes with an open mouth on top. Through these openings the envelopes are filled and finally sealed. Pouches can also be made from two similar or dissimilar composites by effecting a bottom seal in addition to the side seals mentioned above. Alternatively, a pouchlike package can be made in a vertical machine. The web is shaped into a tube, a back seam is made, the bottom of the tube is sealed and product placed into the bag. Finally the upper end of the bag is sealed, closing the filled bag and simultaneously effecting a bottom seal for the next bag.

Formed Packages

A flexible or rigid plastic composite can be formed into any desirable shape. Forming techniques are summarized and illustrated in Figure 10-1.

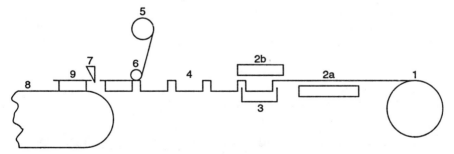

Figure 10-1. Form/fill/seal equipment schematic.

FILL

Placing the product into the package may range from the hand loading of ham into a preformed pouch all the way to the fully automated loading of franks into a thermoformed tray.

SEAL

The four most popular sealing mechanisms are described here.

Bar Sealer. A jaw-type bar sealer with heat on one or both jaws is thermostatically controlled. Both pressure and dwell time may be manually or automatically controlled. Resilient lower jaw permits sealing over wrinkles.

Rotary Bar Sealer. A rotary bar sealer may have optional preheater. One roller could be resilient. Either or both rollers could be serrated to give crimped seal.

Impulse Sealer. This type of sealer has cold jaws. Heat is furnished by a resistance wire that gets an impulse of current at time of sealing, only.

Banded Sealer. Endless metal bands carry material between heated jaws, pressure rolls, and cooling jaws. The thin bands transmit heat rapidly, hold material in contact while heating to form seal and cooling to set seal.

THERMOFORM

Many materials can be formed by heat and pressure into predetermined shapes. Nylon in combination with a polyolefin gives the best results at a reasonable cost. Some people use polyester in place of nylon. This, however, limits the depth of draw severely and may result in package failure. Nylon in addition to thermoformability has good all-around strength characteristics. This combination provides for excellent package performance from the filling station to the ultimate user. Polypropylene, too, has good forming characteristics and can be used—alone or in conjunction with sealants—for ffs applications. Its use should be limited to heavy gauges of cast film only.

FORM-FILL-SEAL EQUIPMENT

Several excellent machines are available for producing thermoformed plastic packages. They all have certain features in common, outlined in Figure 10–1. An understanding of these features may assist in assuring the proper evaluation of materials. Naturally, many machines have unique gadgetry and specialty features which set them apart from the rest of the pack. However, it is usually the basic mechanical and electrical design that is responsible for the performance of the equipment.

Unwind

The thermoformable material may consist of one or more layers of plastics of from 0.003″ to 0.03″ in thickness. The product is supplied in specified width on 3″ or 6″ core. *Important:* Material order must specify:

   a. Width (with tolerances).
   b. Core size.
   c. Sealant position (in or out).

The film is unwound (position 1 in Figure 10-1) and *clamped* as it enters a heating station. If the equipment lacks clamping one may experience problems with forming film under 0.01″ in thickness.

Heaters

There are two ways of heating the film in order to soften it prior to forming (shaping) it.

   a. Contact heat (2a in Figure 10-1). The film slides over a heated plate. This system is not very suitable for heavy gauge films.
   b. Radiation heat (2b in Figure 10-1). It has many advantages over the contact heater. The heater can reach a substantial temperature and still be cooled rapidly when the machine stops. The film can also be formed while still under the heater.

Forming

The softened film is formed to the shape of a die (3 in the figure) by:
   a. Vacuum, which pulls it into the die.
   b. Pressure, usually in conjunction with vacuum, which assists the vacuum forming.
   c. Male plug, which pushes the film into the die.

Loading

The formed shape is thereafter advanced (4 in the figure). As it cools it is ready to have the product placed into the formed container. The speed of loading may range from 15 to possibly 120 packages a minute. At the lower range, one to four loaders can handle the placement of product. At the high range, automatic loading equipment is required.

Secondary Unwind

The lidding material, with or without printing and made from a variety of materials, is suspended from this unwind position (5 in the figure). *Important:* order must specify:

    a. Width (not necessarily same as forming).
    b. Core size (not necessarily same as forming).
    c. Sealant position (in or out).
    d. Print configuration (FPA number).

Sealing

The lidding material is sealed to the formed web (6 in the figure). The temperature is adjustable over a wide range. However, the temperature indicator is often at wide variance with the actual temperature of the sealer. Furthermore, most of these sealers can maintain a temperature of $\pm$ 25°F, at best. Thus a setting of 250°F is in reality 225 − 275°F. Optionally, the package is evacuated and/or gas filled at the sealing station.

Cutoff

The finished package (9 in the Figure) is now severed. This separation may be as simple as a razor blade slitting, a guillotine cut, or as complex as a die cut (7 in the Figure). A conveyor (8 in the Figure) takes the finished package to an inspection and/or packoff station.

FORMABLE MATERIALS

The use of rigids such as PVC, XT, Barex, polyester, and others has been considered in Q 8. There are an almost infinite number of combinations that can be prepared from Nylon, polypropylene, polyester, polyethylene, EVA, and other thermoformable flexible plastics in laminations or coextrusions. In recent times, resin alloys have gained in popularity as well.

APPLICATIONS

In selecting a thermoformable web one must give consideration to cost effectiveness. It is essential to select a product which will provide good package performance at a reasonable cost. As a package is formed, the material thins, especially in the corners. The degree of thinning is a function of depth of draw, surface area, gauge of film, type of material, forming temperature, and several other factors.

Generally speaking, all other conditions being the same, the package application with the higher draw ratio will perform better. We define draw ratio (D. R.) as

$$D. R. = \text{surface area}/(\text{depth of draw})^3$$

Thus, the larger the surface area and the shallower the draw, the better the chances of utilizing a less expensive packaging material. It is best to designate packaging applications according to forming difficulty encountered as well as the product to be packaged.

Normally it is advantageous to thermoform at the lowest temperature allowable (see Figure 10–2). The residual corner thickness of the thermo-

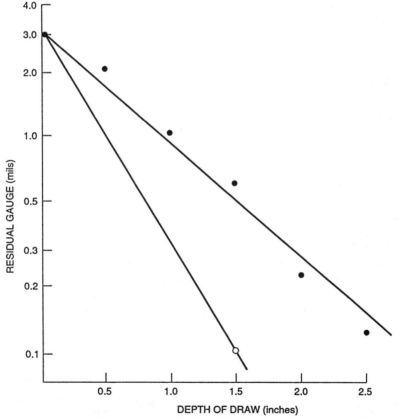

Figure 10–2. Effect of forming temperature on residual thickness. ° = 900°F, * = 1100°F

formed package is substantially better the lower the forming temperature. This is best achieved by lowering the temperature to a point where complete forming is no longer possible, then raising the forming temperature gradually to just a point where adequate forming is noted.

The effect of increased Nylon gauge is shown in Figure 10-3. Even at very shallow draw, the heavier Nylon composite performs significantly better. At deep draw, the increased Nylon gauge is indispensable. The effect of increasing the sealant gauge is shown in Figure 10-4 (Hirsch and Grippenburg 1978).

The data assembled in the unpublished study just cited dealt with many other parameters, as well, and are too extensive to be fully reported in this limited space. However, one may conclude that many factors must be considered in the selection of a thermoformable package.

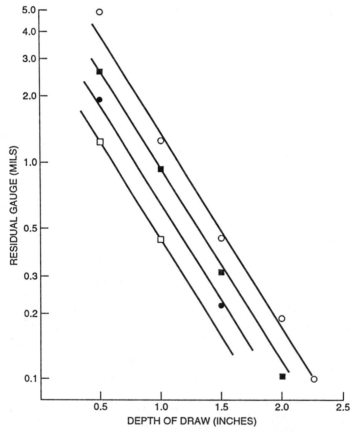

Figure 10-3. Effect of nylon gauge on residual thickness. + = .00075 in., ° = .001 inc., # = .0015 in., * = .002 in.

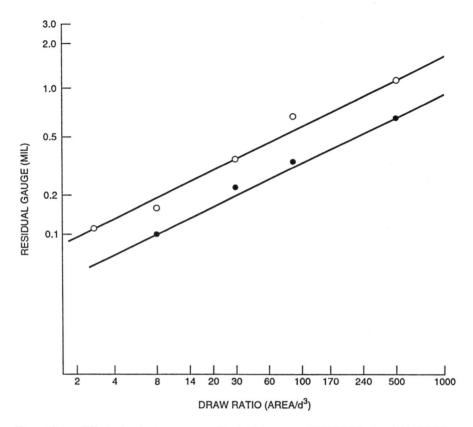

Figure 10-4. Effect of sealant gauge on residual thickness, ● = .002″ LDPE, ° = .003″ LDPE

CASE STUDY 10-1

A regional packer of hot dogs placed his product in premade pouches. The ten franks were positioned in a single layer requiring an oversized bag of 9″ × 13″ to accommodate the product. The operation was labor intensive. Five employees managed to pack about 250 packages per hour.

When this packer introduced ffs equipment, he reaped many benefits. The package size was reduced to about 5.5″ × 6″. Thus the overall material consumption was reduced from 234 SI to 66 SI, a savings of 168 SI per package. The materials for the new package were more costly than those employed for the pouch. However, due to the more than 70% quantity savings in materials, the overall cost of the new package was appreciably lower than that of the pouch.

One additional benefit the processing plant gained from the new equipment was increased productivity. Two employees were able to service the packaging operation, producing 1500 packages per hour. Thus the ffs equipment increased production 600% with 40% of the workforce. The capital investment paid for itself in short order.

ADDITIONAL READING

Anon. 1988. Visual inspection system focuses on total quality control. *Food Engineering* **60**(1): 91–92.

Hirsch, A. and Grippenburg, C. 1978. Thermoforming parameters. Private report.

Hixson, R. W. 1985. Ten-step guide provides an orderly procedure for solving packaging problems. *Industrial Engineering* **17**(2): 86.

Lantos, P. R. 1985. Polymer blends and alloys will upgrade food packaging. *Packaging* **30**(5): 59–62.

Larson, M. 1988. On-line inspection offers simple choices. *Packaging* **33**(1): 68–71.

Lewis, F. C. 1989. Form-fill-seal. *Packaging* **34**(5): 125–127.

Trombly, J. E. 1988. How to specify a machine vision system. *Packaging Technology* **2**(2): 50–51.

Q 11: How Is Aseptic Packaging Accomplished?

A 11: Aseptic means free of pathogenic microorganisms. It is desirable to package food with as low a bacterial population as is feasible. In addition we attempt to retard the further growth of microbes by refrigeration, controlled atmosphere, or freezing. Another approach to food preservation depends on post-packaging thermal treatment to reduce the microbial count. A brief heat treatment (30 minutes at 143°F or 15 sec at 160°F), known as pasteurization, kills most but not all pathogenic microbes. The product thus treated can be retained under refrigeration for several days. A much more severe heat process (15 minutes at 250°F) will achieve a bacteria-free product which can be stored at room temperature for months or even years.

The advantages are apparent. The container of milk which would spoil in hours, can now with the aid of pasteurization survive in a refrigerator for about a week. But why not sterilize the milk and preserve it for much longer? The answer to this question can be found in all the disadvantages of the sterilizing process. The high temperature and long exposure cycle are costly, they alter the flavor, texture, and appearance of the food product. Often the high temperature destroys vitamins or other nutrients. To prevent reentry of microbes and shield the sterile product from damage, the package should be sturdy, such as a tin can. This, however, adds to the cost of the product. Furthermore the weight of the can increases the shipping cost and exacerbates an already serious disposal problem.

Aseptic packaging offers an opportunity to eliminate all the disadvantages of sterile packaging, while retaining all the advantages. In the canning system, the food is packaged in cans, these are sealed, and the finished cans are sterilized in a pressure cooker at about 250°F for a longer time interval.

The time and temperature cycle depend on the product and the loading of the sterilizer. In the aseptic system, the product is sterilized just prior to placement in the container. The plastic container is sterilized in line just prior to loading, and the atmosphere around the packaging operation is sterile and shielded from contamination. Thus the sealed package coming off the packaging line is sterile and requires no refrigeration.

MATERIALS

The packaging material should be lightweight yet sturdy. It must be sterile and prevent bacterial penetration. It should exhibit barrier properties, retarding oxygen and flavor migration. A number of PVDC and EVAL® coextrusions (see Q 6) meet the latter requirements. With regard to sterility, rolls of packaging materials were tested monthly for more than ten years for microbial surface contamination. Not once during these more than 120 tests was a positive result reported. Nevertheless, positive steps are taken to preclude the presence of any microorganisms. The packaging material is treated to kill any surviving microbes on its surface. The most popular method involves a hydrogen peroxide dip. This is followed by a sterile water rinse, a squeegee to remove excess liquid, and infrared/UV drying. The material is now ready to be shaped (thermoformed) or made into pouches. In either case, the packaging material is isolated from the moment it enters the presterilization until it leaves as the finished sealed package.

THE PRODUCT

Not all foods are suitable for this type of packaging. At this writing, the system is limited to some liquids that can be sterilized and delivered easily. Thick slices of ham, in contrast, would present very serious difficulties in either respect. Acidic liquids, having a natural pH of 4.6 or less, are favored because they naturally retard bacterial growth. Thus, orange juice was among the first food products to be aseptically packaged.

The product must be placed in a sterile condition into the presterilized container. To achieve this with minimal damage to the food item, the sterilization takes place at about 290°F for about 3 seconds. The sterile product should be chilled and delivered directly into the package.

THE ENVIRONMENT

The entire system is enclosed and thus shielded from environmental contamination. Air is sterilized at very high temperature (300°C/570°F +) and after cooling maintains a positive pressure throughout the system to prevent nonsterile invasion to compromise the safety of the process. UV lights are placed in strategic positions to enhance the environmental sterility.

APPLICATIONS

In the USA aseptic packaging has made some gains. The major advances, however, have been made in western Europe. The little creamers in U.S. restaurants are polystyrene cups which require refrigeration. In Munich, this same product is in a polypropylene/PVDC/polypropylene thermoformed cup. On the bottom of this cup is an embossed expiration date, and the product requires no refrigeration for weeks or months.

Juices, too, are more readily available in aseptic packages in Europe than in the USA. While there are various size packs produced, the vast majority is in single portion configuration. The same applies to puddings, flavored milk, and soft ice cream.

PROGNOSIS

Old habits are hard to break. The tin can has more than a hundred year success story. The canning operation is speedy—much faster than the aseptic packaging operation. The equipment is in place and the production employees well trained. Thus there is a natural reluctance to invest money in new equipment and train a new crew to switch to the aseptic system. There is no compelling need in North America or in Western Europe to make this change. The can may survive at room temperature and the bottle or flexible package could be kept in the refrigerator for days or weeks. It is in the third world countries that aseptic packaging would be especially appreciated, since refrigeration is not widely accessible. However, the cost of the aseptically packaged product is beyond the reach of this mass market.

Another factor holding back the wider acceptance of aseptic packaging is the very limited range of products offered. What about soups, puree of potatoes or vegetables, or scrambled eggs? If the package were microwavable that, too, would help to further its popularity.

CASE STUDY 11-1

A packer of fresh orange juice had a range of problems starting up an aseptic packaging operation. This is not unusual, since the introduction of any type of new operation is expected to be plagued by all sorts of problems.

The orange juice met all bacteriological requirements, but changed color in a few days. A systematic evaluation of all possible causes provided the answers. Considered were:

a. Excess heat in the sterilization process.
b. Residual peroxide from the presterilization of the web.
c. Excess residual oxygen in head space.
d. Excess air (oxygen) in the juice.
e. Inadequate barrier provided by packaging material.

The problem was resolved by meticulous attention to all details.

ADDITIONAL READING

Anon. 1990. Aseptic low acids find ffs. *Packaging Digest* **27**(6): 90–94.

Anon. 1988. Bosch reveals latest packaging development. *Packaging News* (UK) (1): 11

Anon. 1982. Aseptics: sterile packs with a fertile future. *Packaging Digest* **22**(10): 107–126.

Erickson, G. 1988. Flashy packages reflect new consumer tastes. *Packaging* **33**(1): 58–60.

Folkenberg, J. 1989. Beyond the tin can. *FDA Consumer* **23**(9): 34.

Henyon, D. K. 1988. Aseptic packaging: past, present and future. *Standardization News* **16**(3): 40–43.

Kessel, M. J. 1988. Canning industry: metal vs plastics. *Food Engineering* **60**(1): 87–90.

Kim, P. and Binning, R. 1987. Aseptic technology. *Confructa Studien* **31**(5/6): 126–132.

Knill, B. J. 1987. High-barrier plastics have worldwide appeal. *Food and Drug Packaging* **51**(2): 8, 32.

Mana, J. 1988. Showcase: aseptic packaging. *Prepared Foods* **157**(3): 106–109.

Phillips, B. G. and Miller, W. S. 1973. *Industrial Sterilization*. Durham, NC: Duke University Press.

Sfiligoj, E. 1988. Aseptics: carving out a new identity. *Beverage Industry* **79**(4): 60–61.

Q 12: Are the Product and the Package Compatible?

A 12: Far too frequently little thought is given to the suitability of a package as it relates to the product it is meant to contain. Many questions have dealt with the effect of the package on the product. It is only appropriate to consider the potential effect of the product on the package itself.

APPEARANCE FACTOR

Some packaging taboos are self evident. Everyone knows not to place a wet product in a paper bag. Greasy products should not be wrapped in ordinary paper; the fat stains are just too repulsive. However, many other appearance problems have been overlooked on occasion. The effect of the product on the packaging material is often learned the hard way.

For example, nuts packaged in a cellophane/polyethylene bag transfer oil droplets onto the poly surface. After some motion—and there is plenty of movement during the ordinary shipping cycle—the pouch has oil droplets or smears on the inside surface. Similarly, mustard packs show a distinct yellowing as the product migrates through white polyethylene.

CHEMICAL COMPATIBILITY

There are many products which are difficult to contain. Hydrofluoric acid, for example, will consume a glass bottle in which it is stored. Many plastics are sensitive to high acidity or high alkalinity or both. EVA, EAA, and ionomer are examples of plastics that would be severely altered by products exhibiting pH extremes. Solvents such as alcohols, ketones, or hydrocarbons may destroy the package containing them. Yet alcohols are common ingredients of moist towlettes, scented toiletries, and of course alcoholic beverages. Ketones and hydrocarbons may be present in cleaning products, polishes, and a variety of household items. The solvents may raise havoc with a package in one of two ways. First, most sealants are solvent sensitive. Thus in constant contact with the solvent the seals weaken and the package spills its content. Another mode of failure is induced by the migration of the product through the sealant. The solvent may attack the adhesive and cause delaminative failure of the package. In simple terms, the package falls apart. The solvent may, however, migrate rapidly through the outer walls of the package without attacking the adhesive. The integrity of the package is thus preserved, but the loss of solvent changes the product. Its appearance as well as utility may suffer irrevocable harm.

Essential oils are difficult to contain as well. "The package that smells good sells good!" may be a good marketing slogan. However, aroma emanating from a soap package originates from and is lost to the product. The soap is thus deprived of the perfumes lost through the package walls. In the long run this must hurt the sale of the product.

Similarly, consideration should be given to the packaging of baked products. The aroma emanating from the bakery is most pleasing to the passerby. But the same aroma retained in the baked item would assure many more satisfied customers.

Many other chemicals have become constituents of everyday products. It is no simple matter to devise a package to contain iodine or one of its compounds. Chlorine compounds employed as bleaches or disinfectants are a special source of concern for the packaging technologist.

Many a product has components that attack the packaging material. Unfortunately, the disintegrating forces often act slowly and package failure is not apparent until the product has reached the retail store. A number of well publicized law suits have been brought by food packers against suppliers of packaging materials. The contention has been upheld by the court that there is an implied assurance that the package will indeed contain the product for a reasonable shelf life expectancy. It would reduce litigation and save embarrassment if all parties would agree on definitions prior to package selection.

PHYSICAL ENDURANCE

The selection of the packaging material must also take into consideration the physical stress exerted by the product. Ground coffee, for example, is highly abrasive. Placing this product into a plastic pouch creates unusually tough demands on the endurance of the package. There are of course many such products with sharp edges or corners. Corn chips, frozen vegetables, and noodles cut, tear, and abrade packaging material severely. Puncture can also be a serious problem inherent in the nature of the product. Bone-in meat presents one of the worst puncture problems. Meat packers have elected to avoid this difficulty by placing a "bone guard," a wax-impregnated cloth, over the sharp bone edge.

TESTING

It cannot be too strongly emphasized and must be repeated time and again that there is no substitute for field testing. Even the most astute packaging technologist is often lured into the trap of drawing erroneous conclusions from inadequate data. The packaging function is at times under pressure from marketing to get the new product into the market as rapidly as possible. Shelf stability studies for up to two years are thus out of the question. One must draw conclusions from *accelerated aging* (see below) and project behavior of the package and the product over an extended time period. Under such circumstances, the packaging technologist is utilizing his past experience as a forecast guide. It is important to remember that prior experience is applicable to identical situations only. The fact that nylon/polyethylene is a good packaging material for bacon does not automatically make it equally good for packaging pork chops. Similarly, a printed polyester/ EVA may perform satisfactorily for the packaging of luncheon meat. A change in the printing inks cannot be assumed to be an insignificant alteration in the packaging materials. The experience gained with the prior printed materials is not applicable to the new inks, or the new print design. There have been cases in which the ink change caused residual solvents in very minute trace quantities to migrate into the food and impart an off flavor. It has also occurred that changes in the print design, even with the very same inks, produced undesirable effects. Where in the former design there was just a small printed area in the center portion of the package, the new overall print left ink in the seal area which led to delamination. These few examples are cited here as a warning that some so-called minor changes may possibly have dire consequences.

ACCELERATED AGING

Studies conducted at Standard Packaging Corporation's research laboratories tried to assess the interaction between certain ingredients frequently found in prepared food products and a variety of single-ply packaging materials. Approximately 10 grams of each ingredient was placed in a small pouch and retained at room temperature for a period of four weeks.

Weights were taken weekly and percentage gain or loss recorded. It was assumed that weight gains were primarily due to moisture absorption (and to a much smaller degree due to oxidation) while weight losses represented migration of the product through the packaging material. Results of this study are summarized in Tables 12-1 and 12-2. In addition to the weight changes, packages were also checked for external odor (Table 12-3). This, too, was taken as an indication of active migration of the product through the film. One must remember, though, that some products have no inherent smell and cannot be evaluated by aroma emission. Table 12-4 summarizes the visual inspection of the product as well as pouches on a weekly schedule.

While such information is very useful, it does not supplant the evaluation of the actual proposed package. Product should be placed in the proposed package for storage under conditions as specified below. The number of packages thus prepared will depend on the duration of the test. If at all possible, the number of packages prepared should exceed the minimum number required. During the progress of the evaluation cycle one often thinks of additional tests to be performed or of extending the entire cycle. Preparation of additional samples at a subsequent date may delay the completion of the experiment and place the results in question. Even though the experiment may last only several weeks, one should retain some samples for months or years in order to ascertain the ultimate appearance and performance characteristics of the package and its content beyond the expiration date.

In this test, packages are normally divided into three groups and stored at:

ambient	75°F (24°C)/50% RH
jungle	105°F (40°C)/100% RH
desert	125°F (52°C)/10% RH

Table 12-1. Weight Change through Sealants, 4 Weeks Storage.

	2.0 mil PP, % (Loss) Gain	2.0 mil LDPE, % (Loss) Gain	2.0 mil Surlyn, % (Loss) Gain	2.0 mil EVA, % (Loss) Gain
Butter	(2.9)	(9.4)	(8.3)	(11.9)
Garlic	(4.4)	(7.1)	(11.0)	(14.4)
Lard	.004	.043	.023	.051
Vanilla	(3.7)	(25.0)	(26.8)	(41.2)
Citric acid solution	(2.8)	(6.9)	(6.9)	(16.6)
Ethyl alcohol	(3.6)	(56.6)	(62.9)	(72.0)
Glycerine	.54	1.06	1.14	1.92
Menthol	.01	(5.5)	(.88)	(16.4)
Citric acid powder	(1.5)	(3.0)	(3.2)	(4.9)

Table 12-2.   Weight Change, 4 Weeks Storage.

Product	2 mil HPDPE, % (Loss) Gain	2 mil Rubber Mod. HDPE, % (Loss) Gain	2 mil Vinyl, % (Loss)	2 mil Saranex, % (Loss) Gain	2 mil MDPE, % (Loss) Gain	1.5 mil 22A Aclar, % (Loss) Gain
Butter	(2.34)	(1.58)	(10.5)	2.49	(4.27)	(0.17)
Garlic	(3.16)	(1.33)	(43.4)	(2.02)	(2.40)	(0.50)
Lard	(0.004)	(0.12)	(0.81)	0.043	0.015	(0.076)
Vanilla	(7.73)	(1.51)	(49.6)	(1.58)	(4.58)	(0.15)
Citric acid	(0.72)	(0.52)	(10.39)	(0.84)	(1.28)	(0.081)
Ethyl alcohol	(90.80)	(2.58)	(14.05)	(0.42)	(10.17)	(1.71)
Glycerine	0.39	(0.76)	(1.06)	0.022	(1.57)	(0.028)
Menthol	(0.41)	(0.22)	(2.10)	(0.34)	(0.54)	(0.058)

Table 12-3. Odor Migration (Sniff Test).

Product	HDPE	Rubber HDPE	Vinyl	Saranex	MDPE	Aclar
Butter	No	Yes	Yes	No	Yes	No
Garlic	Yes	Yes	Yes	Yes	Yes	No
Lard	No	Yes	No	Yes	No	No
Vanilla	Yes	No	Yes	No	No	No
Citric acid	No	No	No	No	No	No
Ethyl alcohol	No	No	No	No	No	No
Glycerine	No	No	No	No	No	No
Menthol	No	Yes	Yes	No	No	No

Table 12-4. Product/Pouch Appearance.

Product	HDPE	Rubber HDPE	Vinyl	Saranex	MDPE	Aclar
Butter	1/4	1/4	1/4	1/4	1/4	1/4
Garlic	1/4	1/4	3/4	2/4	1/4	2/4
Lard	1/4	1/4	1/4	1/4	1/4	1/4
Vanilla	1/5	1/5	3/5	1/4	1/4	1/4
Citric acid	1/4	1/4	1/4	1/4	1/4	1/4
Ethyl alcohol	*/5	1/4	1/4	1/4	1/4	1/4
Glycerine	1/4	1/4	1/4	1/4	1/4	1/4
Methanol	1/4	1/4	1/4	1/4	1/4	1/4

1. No change in product appearance.
2. Product discoloration.
3. Product dried; solidified; crystallized.
4. No seal failure.
5. Seal failure; seals attacked by product.
* All alcohol evaporated.

Packages kept under these conditions are examined at periodic intervals. The frequency of tests should be such as to provide 3–4 evaluations during the selected test cycle. Thus, in an accelerated cycle lasting 4 weeks, tests should be performed each week. If the cycle lasts 3 months, tests should be conducted every third week. The number and type of tests to be performed will vary. Evaluation should never be limited to a single package. Many factors may contribute to an exceptional result. One determination could be the exception rather than the norm. Ideally, the test results are the average of a large number of determinations. One must realize that many repetitive tests are costly and time consuming. Nevertheless, a single test result is meaningless, while a duplicate could be acceptable, if the results are replicated. However, since variant results on a duplicate test would still be inconclusive, it is best to elect testing in triplicate as the minimum quantity. Tests most frequently conducted include:

Weight	Loss or gain indicates barrier condition of package. Severe changes may be related to package failure.
Seals	Visual inspection followed by seal strength determination.
Bond Strength	Visual inspection of opened package, observing ply separation or discoloration of packaging material, followed by measurement of force required to separate plies.
Visual Inspection	Both packaging material and product should be examined carefully to note any change whatsoever.

IN CONCLUSION

- Do not take anything for granted.
- Do not assume any change to be of no consequence.
- Do not just hope for the best—do something to assure the packaging success!

CASE STUDY 12-1

The suitability of two packaging materials for containment of a salad dressing was evaluated:

| Laminate 1 | 195 M Cello/Adhes./0.00035″ Al Foil/Adh./0.0015″ PP |
| Laminate 2 | 195 M Cello/10 lb PX/0.00035″ Al Foil/PX/0.0015″ MDPE |

One hundred pouches of each laminate were prepared and filled with the salad dressing. Approximately 33 pouches of each were placed at three storage conditions (ambient, jungle, and desert). All pouches were weighed and visually inspected for obvious flaws on day zero. Ten pouches of each material were withdrawn from each of the three storage conditions each month, for a period of three months. The weights were recorded. The test results are summarized in Tables 12-5 and 12-6. The conclusions are:

1. Environmental conditions caused minimal degradation of the laminates evaluated.
2. Seal strength of the packages decreased slightly after storage at high temperature and high humidity. However, the resultant seals were sufficient to contain the product.
3. No delamination was observed in the sealant to foil layer.
4. As might be expected with foil laminations, package weight changes

Table 12-5. Storage Studies of Salad Dressing Packaged in Laminate 1.

	Storage Conditions					
	Ambient		Jungle		Desert	
Storage Duration	Weight Change (%)	Seal* Strength (lb/1″)	Weight Change (%)	Seal* Strength (lb/1″)	Weight Change (%)	Seal* Strength (lb/1″)
0 week	—	4.7	—	4.7	—	4.7
1 month	+0.54	4.9	+0.16	4.3	−0.42	5.0
2 months	+0.38	5.4	−0.60	5.4	−0.47	5.0
3 months	+0.47	5.6	−0.90	4.2	−0.87	5.0

*ASTM F-88.

Table 12-6. Storage Studies of Salad Dressing Packaged in Laminate 2.

	Storage Conditions					
	Ambient		Jungle		Desert	
Storage Duration	Weight Change (%)	Seal* Strength (lb/1″)	Weight Change (%)	Seal* Strength (lb/1″)	Weight Change (%)	Seal* Strength (lb/1″)
0 week	—	6.2	—	6.2	—	6.2
1 month	+0.13	6.4	+0.40	4.6	−0.77	4.8
2 months	+0.94	7.5	−0.70	5.9	−0.60	6.4
3 months	+0.42	5.3	−0.96	5.2	−0.90	5.6

*ASTM F-88.

were slight over the 3 month storage period. In all cases this amounted to less than 1%.

ADDITIONAL READING

Anon. 1988. Resealability reigns in flexibles. *Packaging Digest* **25**(4): 42–48.

Anon. 1985. More convenience in food packaging. *Futurist* **19**(4): 60.

Anon. 1975. Chemical resistance of many common plastics. *Plastic World* **33**(2): 27.

Cole, P. 1988. Focus on flexibles: there's no stopping plastics. *Canadian Packaging* **41**(4): 22–27.

Erickson, G. 1990. New trends make food a challenging game. *Packaging* **35**(1): 44–48.

Henning, H. F. et al. 1976. Toxicity of polymers and polymer raw materials. *Chemistry and Industry (UK)* **11**(6/5): 463–473.

Hu, K. H. and Breyer, J. B. 1972. Pinhole resistance of flexibles. *Modern Packaging* **45**(1): 47–49.

McCarron, R. M. 1972. Package that smells isn't package that sells. *Candy and Snack Industry* **137**(10): 54.

Paris, E. 1984. Packaging (plastic upsurge). *Forbes* **133**(1): 224–226.

Pintauro, N. D. and Simha, A. H. 1976. Teamwork: your key to longer shelf life. *Package Engineering* **21**(4): 38–40.

Shaw, F. B. 1977. Toxicological considerations in the selection of flexible packaging materials for foodstuffs. *Journal of Food Protection* **40**(1): 65–68.

Tzouwara-Karayanni, S. et al. 1987. Adsorption of vinyl chloride onto plasticized PVC by classical partition in the presence of various food simulating solvents. *Lebensmittel Wissenschaft und Technologie* **20**(4): 202–206.

Q 13: Do Government Regulations Control Package Selection?

A 13: There are laws, rules, and regulations to cover every aspect of the packaging industry. By the time a package reaches a consumer it has been scrutinized by an army of government regulators. Compliance with all the federal, state, and local laws has become so complex as to require legal counsel. It is sheer folly to attempt to navigate through the morass of legislation on one's own—or even with the aid of a generalist law practitioner. It takes a specialist, who knows his way around the regulatory agencies, to survive this mess.

It is impossible to impart all this complex knowledge in a few paragraphs—or even in a single volume devoted to this subject. What is offered here is an overview with a few examples. It should be remembered that ignorance is no excuse for violating the law. In addition to severe fines, some offenses may actually lead to jail terms. In addition, failure to observe the rules may spawn consumer complaints, with consequent liability law suits.

FDA

The Food and Drug Administration has sweeping powers to regulate food and its packaging. Originally, labeling was its major focus of concern. The term "label" is not restricted to a tag attached to the package—it also includes all information printed directly on the package. The size of print, the quantity measures, and the ingredients must appear in a specified manner. The FDA was and is also very concerned about unwarranted health claims for foods. New rules drafted in late 1989 prohibit all claims that any food product promotes health—except those pointing out a link between:

1. Low fat content and reduced risk of heart disease.
2. Low fat content and reduced risk of cancer.
3. High-fiber diet and reduced risk of heart disease.
4. High-fiber diet and reduced risk of colon cancer.
5. Low-salt diet and prevention of high blood pressure.
6. Calcium-rich foods and prevention of osteoporosis.

These six claims on packages or in advertisements will not be challenged. Any other health claims for foods will be allowed if adequate clinical proof can be presented. However, such evidence will be very critically scrutinized.

Some local packer may delude himself into believing that FDA rules are not applicable to local business establishments. Just one ad in a local paper which happens to have an out of state subscriber places him in interstate commerce and opens the door to FDA intervention. A consumer carrying a package from a local store to another state may trigger a similar event. In any case, each establishment, large or small, local, regional, or national in scope, is subject to inspection by local and state health authorities as well. Any one of these government authorities can issue summonses, impose fines, and in very drastic cases close the place down.

GMPs

FDA in the 1960s promulgated Good Manufacturing Practices (GMP). The basic idea was not to tell the manufacturer how to run his own business. Rather FDA attempted to encourage the entrepreneur to define his process and commit basic operating steps to writing and police himself. If, for example, a packer of cookies wrote on his GMP statement that each and every single package of cookies would be weighed, then he had better do it. The enterprise cannot just decide to weigh every tenth package and escape an inspection citation. FDA is not arguing that the weighing of each package is essential. The manufacturer is called to task for cheating on his own GMPs. He failed to do what he alleged he was doing.

GMPs are not just installed and adhered to in order to satisfy government inspectors. They assure the production of a satisfactory product and guarantee the safety of employees and consumers. Quality control procedures are spelled out and traceability of each item back from the retail store to the source of all ingredients is covered. A recall, should it be needed, can thus be localized and the damage contained.

Adulterants

The FDA has also clamped down on the presence of unsafe ingredients in processed food. This involves a long list of items which may not be added into the food as preservatives, colorants, or humectants. The Delaney clause has mandated the exclusion of items that are carcinogens (cancer causing), and this has been a source of prolonged debate. Should ingredients be banned because they seem to cause cancer in laboratory rats when ingested in massive doses?

An actual example comes to mind to illustrate this point. The FDA was about to ban a component of an ink employed to print a flexible bacon package. The ink was not in direct contact with the meat, but the concern was expressed that a carcinogen could migrate through the plastic and be absorbed by the bacon. A presentation before FDA argued that figures on the migration rate of the chemical in question were not available and would be costly and time consuming to obtain. However, if one assumed that one hundred percent of the chemical in the ink was absorbed into the bacon and if a grown man were to eat one pound of bacon daily for the next seventy years, and if all this chemical were to accumulate in this person's body—then and only then would a lethal dose have been consumed. FDA admitted the absurdity of all these suppositions and relented in this specific instance. However, they have been adamant on other occasions and prohibited other ingredients on the flimsiest of suspicions. An elaborate scheme of approval or disallowance has evolved which classifies materials for direct or indirect contact. The supplier of such items will certify their permissable use.

USDA

Meat, poultry, and eggs are subject to U.S. Department of Agriculture inspection as well. Thus a manufacturer of frankfurters may have to abide by FDA and USDA regulations. USDA is very concerned about the seller cheating the consumer with either false weights or erroneous impressions of freshness. For example, it was quite common to print red lines in a diamond pattern on a frank package. USDA felt that this practice made the franks look redder than they really were and prohibited this type of package.

OSHA

The packer, having dealt with FDA, USDA, and local and/or state health authorities, must face all the many safety requirements in the workplace. Some of the packaging operations can be dangerous, with exposure to cutting, grinding, baking, and similar equipment. There can also be a serious problem with electric wires and connections on wet floors and workbenches.

In recent times right-to-know legislation has mandated training of personnel and ready access to information regarding potential hazards associated with all ingredients present in an establishment. These data sheets must be made available to the community as well, to forestall accidentally created hazardous situations.

EPA

A packaging operation must deal with its potential impact on the environment. Provisions must be made to limit air pollution. Normally, one thinks of black smoke rising from factory stacks, but there are many other ways of fouling the air. It has been said that the "U.S. industry is using the sky as a garbage dump, disposing of 360 million pounds of suspected cancer causing chemicals in the air annually." While this statement's veracity may be questioned, it was at least qualitatively descriptive of the sad status of air quality. EPA is likely to restrict emissions further. For example, in November of 1989 new trash burning rules were formulated which will require recycling of 25% and thus reduce emissions by a high percentage. Odor is no longer an unavoidable nuisance. The unbearable aroma emanating from the Chicago stockyards in the 1930s would not be tolerated today. The EPA is empowered to levy heavy fines and if need be to shut an operation down for endangering the health of the neighboring population.

Disposal of effluents into streams, rivers, lakes, and the ocean are regulated by EPA as well. In right-to-know regulations it is made clear that personnel must be trained to deal with accidental spills. One or more persons must be designated who will be in charge of such a cleanup. The oil companies in 1989–90 spilled millions of gallons of crude in Alaska, Texas, and in New York. They have been dealt with less harshly than a bakery operation might be for spilling some whey down the drain. One may assume that EPA considers the impact on the nation's economy a shut down of either enterprise may have. It might be best to accumulate liquid waste in 55 gallon drums for disposal at a later date.

Waste Disposal

What was a rather simple matter has turned into a new science with all sorts of legal implications. Businesses have been driven into Chapter 11 by hindsight recognition that some innocuous product was toxic after all. Johns Manville had no idea that asbestos was potentially hazardous. When this became known, JM stopped all production of the product. However, in no way could JM compensate all who ever came in contact with asbestos or pay for the removal of all insulation. Similarly, all who disposed of materials in a very responsible and then legal fashion are now decades later asked to foot the cleanup bill. This cost is estimated at one billion dollars for fiscal 1990.

This dangerous concept has not been challenged in the courts, since no one wants to appear to be against a clean environment. Once upon a time it was unthinkable to be against "apple pie and motherhood." In the 1990s one can be against motherhood, but the new sacred cow is the Environment. Furthermore, the cost of legal challenges excludes all but the few biggest companies, and they are more concerned with their image than with the tax deductable cost of the clean up.

Little is written about noise pollution. However, within a plant OSHA can intervene if the noise level rises above 90 decibels and thus endangers workers. Outside the plant neighbors can obtain relief from local or EPA authorities for excessive noise.

Some of the alphabet soup one encounters in regulatory compliance is listed below:

FDA   Acts under authority of the Federal Food Drug and Cosmetics Act.
USDA   U.S. Department of Agriculture, supervises the processing of meat, poultry and eggs.
USEPA   Recently elevated to cabinet level, this department enjoys broad powers to protect the environment.
AHERA   Asbestos Hazard Emergency Response Act.
CAA   Clean Air Act. In 1988, EPA charged 97 defendants with criminal offenses.
CWA   Clean Water Act.
CERCLA   Comprehensive Environmental Response, Compensation and Liability Act—part of the Superfund.
EPCRA   Emergency Planning and Community Right-to-Know Act.
MSDS   Material Safety Data Sheet. Required for all hazardous substances.
RCRA   Resources Conservation and Recovery Act.
SARA   Superfund Amendments and Reauthorization Act.
TSCA   Toxic Substances Control Act. More than 66,000 chemicals are identified as toxic.

CASE STUDY 13-1

A plant laminating and printing packaging materials had difficulties obtaining a permanent operating permit from EPA. This new plant had installed state of the art thermal oxidizer equipment to burn off all evaporating organic solvents. The system was approved by EPA from the conception through its installation and some minor improvements during operation. Nevertheless, neighbors complained of odors and EPA capitulated to this pressure. The fact that a sewage treatment plant within about one mile emitted a foul odor and several other plants in the immediate vicinity operated under grandfather clause, without the benefit of any emission controls, was not a factor. Neighbors complained that odors were eminating from the new plant and EPA knew better, but found it politically more acceptable to side with the voters rather than a business.

CASE STUDY 13-2

A nationally recognized leader in portion pack jams faced a serious problem. The supplier of lidding material had a slipup in QC procedures and shipped some laminate with residual solvent. The packer, too, failed to follow proper QC protocol and neither tested samples of incoming lidding material nor finished packages of jam. Days later, by some coincidence, the solvent odor was discovered and a frantic scramble to contain the damage ensued. Luckily all the suspect production was still in warehouses belonging to the manufacturer. Thus the recall was purely internal, involving neither distributors nor retailers. Traceability of all batches enabled QC to account for all suspect units. Without such traceability, all existing product units in the universe would have had to be recalled. Such a procedure would have been very costly and damaging to the reputation of a highly respected company.

ADDITIONAL READING

Anon. 1988. Regulatory status report. *Food and Drug Packaging* **52**(5): 8.

Anon. 1987. Label mix ups create new worries for FDA. *Drugs and Cosmetic Industry* **141**(6): 37, 60.

Anon. 1987. Nutritional labeling: the not-so-small print. *FIPP* (UK) **9**(11): 154–157.

Best, D. 1988. Food labeling: trends, truth, and the cost of compliance. *Prepared Foods* **157**(2): 106–110.

Brown, L. J. 1988. What food labels tell you. *Good Housekeeping* **207**(8): 137.

Densford, L. 1988. Committee to assess food safety criteria. *Food and Drug Packaging* **52**(4): 38.

Densford, L. 1987. FDA proposes standard for cholesterol claims. *Food and Drug Packaging* **51**(2): 3, 22.

Goerth, Ch. R. 1987. The legal impact: SARA sings a siren song. *Packaging Digest* **24**(4): 31–32.

Hathaway, J. S. 1988. The Delaney clause and carcinogenic pesticides. *Environment* **30**(11): 4–5.

Hunter, B. T. 1985. Better food labeling. *Consumer Research Magazine* **68**(4): 8–9.

Kaplan, M. M. 1989. Hazardous waste issues drawing federal attention. *TAPPI Journal* **72**(6): 65–68.

Norman, C. 1988. EPA sets new policy on pesticide cancer risk. *Science* **242**(10/21): 366–367.

Sloan, I. J. *Environment and the Law.* Dobbs Ferry, NY: Oceana Publications Inc.

Tonnessen, D. 1988. The label debate. *Health* **20**(7): 65–68.

Q 14: How Does the Bar Code Affect Package Design?

A 14: The Universal Product Code (UPC) is one of the revolutionary innovations of the second half of the Twentieth Century. There are many wasteful and costly operations in retailing. Pricing of each item individually is not only costly and time consuming but may contribute to reduction of the quantity of merchandise on display. The product is exposed to possible damage during pricing as well. Similarly, checkout at the cashier's counter is a major bottleneck. Whoever has waited in line at a supermarket or department store cashier has tasted this annoying imperfection in our modern retailing establishment. The merchant is further concerned with the intentional or accidental giveaway at the cash register. Mistakes of one sort or another are common and may cost the store thousands of dollars.

A newer system attempts to do away with both of these problems. In the process, other benefits have been realized as well. The system employs a code identifying the manufacturer and the item. Since the code for the manufacturer consists of a six-digit number, it allows for up to 1,000,000 suppliers. The code furthermore provides for a five digit number, identifying the individual product. Thus each of a potential one million manufacturers could merchandize up to one hundred thousand items. The modern supermarket could carry one hundred billion distinct items—all handled by the currently instituted bar code system.

The numerical code is not carried in the traditionally legible form. The numbers have been translated into dark and light spaces which can be read by an electronic readout device. If the code appears on the package as it reaches the supermarket there is no need to price each package individually.

Its code identifies the package fully. A price can be posted on display shelves and packages can be placed or replaced on this shelf rapidly. Any change in price (up or down) can be effected by a single alteration of the tag displayed on the shelf.

When the customer presents the selected item to the cashier the code of each item is passed over the electronic detector. The code is fed to a computer which performs several operations simultaneously. The one of immediate concern to the customer is a very rapid checkout. The computer finally presents the total cost of the shopping list and calculates the change due. But it also keeps track of specials, discounts, etc. While the UPC in conjunction with the computer sped up the checkout, it has also provided a series of other benefits. As the computer furnishes pricing to the cash register, it deducts the item from the open shelf inventory and the stock record. The store manager can be alerted by the computer to the need for restocking a certain shelf or reordering a specific item. Pilferage—which is a most serious problem in our large cities—is more easily detected and documented. Since the computer can give an item by item sales volume account on any chosen time interval, it enables management to shift space allotments as needed. The data thus gathered furnishes information on fast or slow moving items more rapidly than would otherwise be available. The system is invaluable in many other ways as a decision making tool.

Now that we understand the why and wherefore of UPC—how does it relate to packaging? It is quite apparent that the bar code symbol must be on the package when it gets into the store in order to derive maximum benefits thereof. In the early experimental days of UPC, the symbols were applied as preprinted stickers in the store. Today the majority of products have the bar code printed on the package.

Several important facts must be observed. Figure 14–1 portrays the standard UPC symbol.

THE PRINTABILITY GAUGE

To establish the printability range of the UPC symbol for a given set of conditions on a printed package or overwrap it is necessary for a converter to verify printability of the symbol, using a printability gauge. This is actually a printing test, preferably incorporated into a regular production run, using the adopted standard printability gauge. This printing test provides a visual check for the customer, the converter and the authorized photoengraver, concerning the point at which the printed lines and spacing deteriorate to make readability difficult under normal printing conditions, conducting regular production runs. The printability rating range, selected

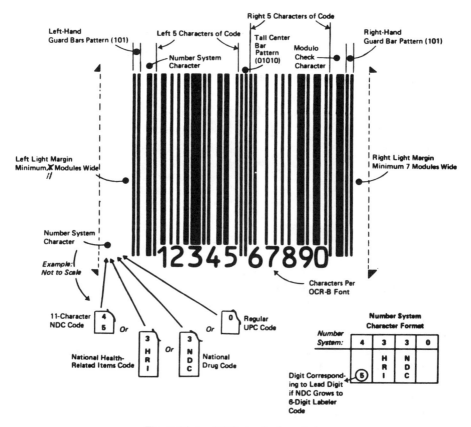

Figure 14-1. UPC standard symbol.

from a point on the printed gauge above the ranges subject to print deterioration and erratic spacing, will provide a means of selecting the proper symbol size and the magnification factor required for the method of package printing being used. The actual printing of this printability gauge, wherever possible, should be either in or close to the location selected for the actual bar code symbol. Printing of the gauge and the UPC symbol on a package are quite similar. Printability data obtained from the gauge printing in the designated symbol area limits the necessity for later symbol size and space requirements changes. A printability gauge (Figure 14-2) is composed of parallel lines 0.0100″ wide, spaced at A and A′ 0.0200″, at B and B′ 0.0180″, at C and C′ 0.0160″, at D and D′ 0.0140″, at E and E′ 0.0120″, at F and F′ 0.0100″, at G and G′ 0.0080″, at H and H′ 0.0060″, at I and I′ 0.0040″, at J and J′ 0.0020″, and at K and K′ 0.0010″.

Following the production run, the converter will evaluate at least 20 sam-

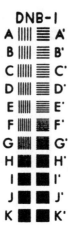

Figure 14–2. Printability gauge.

ple impression sheets, to determine the overall printability level. These samples are usually taken throughout the production run and are quite representative of the printing results. Examination of these results by the converter establishes the printability range. From this information the symbol size required can be determined. These samples are made part of the converter and engraver files for future reference as needed.

There exists a gauge for round containers. To utilize this properly, place the can's diameter in contact with the center point at the center line. This will determine whether the symbol will be scanned properly. It must not exceed 30° when the symbol is centered on the center point. The symbol should be oriented on the label to maximize printability by having the parallel bars oriented in the press direction. The following table shows the absolute limit of 34° for various container diameters as a function of maximum magnification factor.

Maximum Magnification Factor

Container-Diameter (inches)	Full-Size Symbols	Zero Suppression
1.0	—	0.90
2.0	1.00	1.80
2.5	1.30	2.00
3.0	1.54	2.00
3.5	1.90	2.00
4.0	2.00	2.00
4.0+	2.00	2.00

ADDITIONAL READING

Goldberg, G. 1981. Bar codes are here to stay. *TAPPI Journal* **64**(10): 65–67.

Jones, R. W. 1973. How to use the UPC printability gauge. *Modern Packaging* **46**(9): 111–116.

Kelsey, R. J. 1988. Tracking packages in-plant by bar code. *Food and Drug Packaging* **52**(2): 10–14.

Landsdale, D. 1975. Tools of the trade. UPC comparator scale for checking quality. *Package Development* **5**((1) 21–25.

Momiroff, B. C. 1974. Production quality control of UPC symbol. *Package Development* **4**(5): 111–113.

Smith, N. 1974. Application of UPC to flexible films. *Modern Packaging* **47**(10): 311–317.

Q 15: What Is Meant by Nutritional Labeling?

A 15: It has been well documented that what we eat determines the quality and length of our lives. The wrong foods clog our arteries and are in some way shortening our lives. It is thus important to select both the quality and quantity of nutrients for consumption.

The U.S. government has mandated that each package of food, drugs, and lately also cosmetics, disclose a wealth of information about the product contained therein. The White House Conference on food, nutrition, and health in late 1969 recommended that FDA consider the development of a system for identifying the nutritional quality of food. Several subsequent studies have shown that large portions of the population (and not just the poor) suffer from one or another nutritional deficiency. After lengthy study, the nutritional label regulations were published in the *Federal Register* on March 30, 1972. During the early days of this development there were two systems of reporting the nutritional value of food products. The minimum daily requirements (MDR) were one favorite way of expressing nutritional information. However, in recent times the accepted system has become the U.S. recommended daily allowances (RDA). FDA has published a list of RDAs (Table 15-1) which is based on the needs of the average adult male. The RDAs on the package are expressed in percent. The consumer thus avoids the need for memorizing or consulting tables of what the actual requirements are. By looking at the label, the consumer can ascertain that the packaged product may furnish a given percentage of the daily needs of certain vitamins or minerals. By adding up the various items on the menu one should total up to 100% of daily needs.

Changes in RDAs have been published in 1989 by the National Academy of Science. Several basic changes should be noted.

**Table 15-1. U.S. Recommended Daily Dietary Allowance,
Average Male, Age 23-50 (Revised 1980).**

Calcium	800	mg	Thiamine	1.4	mg
Folacin	400	mcg	Vitamin A	1000	mcg
Iodine	150	mcg	Vitamin B_6	2.2	mg
Iron	10	mg	Vitamin B_{12}	3	mg
Magnesium	350	mg	Vitamin C	60	mg
Niacin	18	mg	Vitamin D	5	mcg
Phosphorus	800	mg	Vitamin E	10	mg
Protein	56	g	Zinc	15	mg
Riboflavin	1.6 mg				

Reference Individuals

Age and sex groups have been established by median height and weight as promulgated by the National Center for Disease control and not on some ideal model basis.

Age Groups

Young adults are defined as 19–24 years of age. This allows for higher calcium, phosphorus, and vitamin D values for this age group with continued bone growth.

Infants

Recommendations are based on an equivalent nutrient value of about one liter of human milk (this includes a 25% safety factor).

Vitamin K and Selenium

RDAs have been established for these two products, based on recent nutrition studies.

Reduction in Values

RDAs for folic acid, vitamin B_6, and vitamin B_{12} were reduced by 25–50%.

Zinc and iron were reduced for women only. Thiamin, riboflavin, and niacin, which are involved in energy metabolism, were reduced in relation to the new reference individuals.

The packaged product must carry several additional pieces of information on its nutritional label. First of all, it must state the quantity of product in the package, the serving size, and the number of servings per container. All other information is given on a per serving basis. Furthermore, the information must furnish macro-nutritional guidance. It must state the number of calories per serving and the amount of protein, carbohydrate, and fat per serving. No RDAs for these have been established. The consumer must understand that each person's requirements are based on gender, weight, activity, and many other factors. However, the fact that such information is indeed available to the consumer permits the selection of foods best suitable for daily requirements. When the food item contains less than 2% of the RDA, the statement is not required.

There are many exceptions to the above rule. Certain foods are not labeled in accordance with RDAs. Such foods include infant, baby, and special prescribed dietary foods. Other foods totally excluded from nutritional labeling are those shipped in bulk for either further processing and/or for institutional feeding.

FDA has also clamped down on unfounded claims for nutritional value of foods. It is thus not permitted to say or imply whether in print or through radio and TV advertising that:

1. A balanced diet of ordinary food cannot supply the required amount of nutrients (this has often been claimed in order to promote vitamin supplements or special health foods).
2. That something such as storage, transportation, cooking or other means of handling or preparation has produced deficiencies in the daily diet.
3. That a certain vitamin is superior to another brand (unless scientific data can actually substantiate such a claim).
4. That a certain food can prevent or cure any disease (see six exceptions listed in Q 13).
5. That a food grown on a certain soil is either deficient in or superior in providing certain nutrients.
6. A certain food contains nutrients, when such substances are of no known significant value.

It must be repeated that the above prohibitions are just against unfounded claims of such or similar nature. Where scientific evidence is indeed available to prove the exceptional value of a given food even within the above parameters such may certainly be brought to the attention of the public.

A major policy change is due in the early 1990s. Secretary of Health and Human Services, Louis W. Sullivan, in a speech on March 7, 1990, gave a preview of the first change in nutritional labeling in seventeen years. Until now, nutritional labeling has been required only on food packages that made some nutritional claims. Thus, as long as the bologna package was not advertised as "lean" or as "vitamin rich," there was no legal reason to place a nutritional label on this product. However, the competitive situation in the marketplace often compelled the display of this information on the product. This in turn lead to many abusive practices. The serving size was arbitrarily chosen. A smaller serving, after all, contains fewer calories, or less fat, etc.

The new regulations will apply to all packaged foods, except those with minimal nutritional value such as coffee, tea, spices, etc. The information on the label will have to disclose, in addition to the items mentioned above, cholesterol, fiber, saturated fat, and percentage of calories supplied by fat. There will be efforts to standardize serving size and define terms such as "light," "lean," "low," and similar advertising slogans.

For a while there were plans by the American Heart Association for a new labeling plan called "Heart Guide." Controversy surrounding this idea caused its cancellation in early 1990.

Does nutritional labeling work? Most Americans are well fed. Yet there is an appreciable portion of the population which consumes quantities of food that are not necessarily nutritionally sound. Americans are reknowned for consuming junk food, eating tons of sweets, being on the one hand overweight and on the other undernourished. To date nutritional labeling does not seem to have done the job for which it was designed. Most consumers ignore it entirely. One can station oneself at a checkout counter at any of the large supermarkets and observe the average consumer purchasing a variety of fad foods which have little nutritional value and are relatively expensive.

There was quite a to-do about the TV promotion of carbohydrate-rich fad foods directed at youngsters. Children's programs advertised candies, sugary breakfast cereals, and similar products. Nutritional labeling certainly has not discouraged parents from feeding such foods to their offspring. Placing additional warning labels on advertisements or even on the package will probably have little impact on the consumption of same. Government must accept that it cannot legislate reasonableness.

Nutritional labeling is far more complex than most consumers realize. The regulations have failed to take into consideration that not all proteins are alike and of equivalent value. (Some effort has been made to allow for the presence of essential amino acids). Not all fats are alike—there are those with higher or lower degree of unsaturation—and this too will be incorporated into the labeling system. Finally, the nutritional label does not take quality into account. Two products with identical nutritional labels may prove to be entirely different. One may be barely edible, the other delicious. The nutritional label is incapable of distinguishing this important aspect of food.

ADDITIONAL READING

Alsmeyer, R. H. 1974. Standards and labeling of meat and poultry products. *Association of Food and Drug Officials of the U.S., Quarterly Bulletin* **38**(1): 63–67.

Aulik, D. J. 1975. Nutritional labeling: a challenge for the snack food industry. *Snack Food* **64**(4): 40–41.

Anon. 1986. Congress chews on food-label reform. *Consumer Reports* **51**(3): 142.

Beloian, A. 1973. Nutritional labels: a great leap forward. *FDA Consumer* **7**(9): 10–16.

Brown, L. J. 1988. What food labels tell you. *Good Housekeeping* **207**(8): 137.

Daly, P. 1976. The response of consumers to nutrition labeling. *The Journal of Consumer Affairs* **10**(2): 170–178.

Densford, L. 1987. FDA proposes standards for cholesterol claims. *Food and Drug Packaging* **51**(2): 3–22.

Gutrie, H. A. 1990. Recommended dietary allowances 1989. *Nutrition Today* **24**(1): 43–45.

Harper, A. H. and Lachance, P. A. What's happened to nutritional labeling? *The Professional Nutritionist* **9**(3): 8–13.

Lecos, C. 1988. Food labels: test your food label knowledge. *FDA Consumer* **22**(3): 16–21.

Stern, J. S. 1988. Decoding food labels. *Vogue* **178**(10): 344.

Tonnessen, D. 1988. The label debate. *Health* **20**(7): 65–68.

Q 16: Do Flexible Packages Contribute to Environmental Contamination?

A 16: YES . . . BUT! It is undeniable that we have a solid waste problem. Municipal solid waste has risen from 87.5 million tons in 1960 to 157.7 million tons in 1986. The composition of the waste (see Table 16-1) has changed somewhat over the past 30 years—but not as drastically as the quantity increase might indicate. Paper and paperboard has increased by about 20% in the waste makeup. Plastics have increased almost 15-fold in the same 30-year period. All other components of solid waste have actually declined. One could make a case for paper and plastic packaging having contributed to the increase in solid waste, while cans and glass bottles have diminished in the mix. However, statistics on sanitary food containers are inconclusive in either supporting or contradicting this contention. Unfortunately, the U. S. Department of Commerce has changed classifications in 1987 and thus "paper, coated and laminated, packaging" was established as SIC 2671. Since it was part of SIC 2641, the growth of these items cannot be traced back over the past thirty years. However, in the period from 1987–1990 the increase in packaging materials from 2.2 billion to 2.3 billion dollars is certainly not an untoward rate of growth.

There is another problem peculiar to the U.S. scene. The litter in the streets and on the highways is largely composed of cans, bottles, and other packaging materials. This is not the case in the cities and on the roads of Europe. The cause of litter is thus not the product but the public attitude. However, the same person who discards the food wrapper in the street will then sign a petition condeming the food industry for creating waste. One

Table 16-1.   **Generation of Select Materials in Municipal Solid Waste (Million Tons).**

	1960	1970	1980	1985
Paper and paper products	29.8	43.9	53.9	61.7
Ferrous metals	9.9	12.6	11.6	10.7
Aluminum	0.4	0.9	1.8	2.3
Glass	6.5	12.7	14.9	13.2
Plastics	0.4	3.0	7.6	9.8

may argue that the remedy—eliminating packaging—would be much worse than the environmental contamination (see Q 1). But some relief from the pollution scourge must be found. It can take several approaches, as outlined here.

SOURCE REDUCTION

The less waste we generate—the less we need to dispose. Thus one should reflect at each step of the packaging operation whether *less* will do. Could a box or pouch or wrap be eliminated without hurting the product or its distribution? Could the package size be reduced and yet contain the same quantity of product? A 10% reduction in packaging material use would diminish the solid waste by about sixteen million tons per annum.

INCINERATION

The process has been tried with good results, but is opposed by environmentalists. Garbage can be burnt and the heat generated can be utilized to produce power. Thus the process is profitable and leaves a minimal quantity of ash for burial at dump sites. Environmental concerns about CO_2 generation and the greenhouse effect, hydrogen chloride generation and acid rain, concentrated heavy metals in the ash residue and potential well poisoning, have all combined to prevent this solution from receiving a fair trial.

Some blame must also be attributed to the "not in my backyard" syndrome. All of us want more jails, more and better drug treatment centers, more infectious disease treatment clinics, better mental health hospitals and waste disposal facilities. But all of these—"not in my backyard!"

RECYCLE

The idea of recycling is not as new as some would make us believe. It was quite normal for a mother to recycle cloth diapers. It was equally common for a family to utilize a set of dishes and cutlery for 20 years and thus save about 1.5 million paper plates and an equal number of plastic eating utensils. This sort of waste reduction at the source is much easier and more effective. But this is not what ecologists mean by recycling. They would like the consumer to return the used package to its source, to be reconditioned time and again. With paper for newsprint this is rather simple. Used paper is gathered and returned to the paper mill, where it is dissolved, cleaned, and eventually mixed with virgin pulp to generate recycled paper. The problem with plastics and metals is much more complex.

Consider the simpler situation, that of recycled paper, first. Basic to recycling are five steps: (1) segregation of the waste stream, (2) collection, (3) transportation, (4) reprocessing, and (5) reuse.

Segregation

The consumer who is unwilling to place a few dishes in the dishwasher and would rather pay extra for paper plates is expected to take the time and trouble to separate the garbage. In some states a token deposit is collected on bottles to induce the shopper to return these containers to collection stations. Even the kids do not find it worth their while to pick up bottles off the street for 5¢ a piece. If deposits were $1.00 each, some cleanup effort would become apparent.

Collection and Transportation

The cost of collecting recyclable materials from millions of households and transporting them to a handful of manufacturing facilities for reprocessing is very high. The consumer must bear the cost eventually. Can offshore facilities, unhampered by EPA regulations, produce paper, plastics, and other materials so cheaply as to drive domestic manufacturers out of business?

Reprocessing

Glass milk bottles were routinely reprocessed. They were washed and steam cleaned prior to refilling and going out to retail customers. But many food containers are not suitable for reuse as is. One would not want one's steak wrapped in old newspapers. Yet the old dirty printed paper, mixed with all other recyclable paper, lands in the same vat, where it is deinked, bleached, and in other ways treated to yield a fairly respectable looking paper. There is no objection to utilizing such recycled stock for newsprint or some other commercial application. However, for most food and health care applications a clean, virgin pulp should be compulsory.

Reuse

Most applications do not permit a reuse of a container such as the milk bottle. The old used container must be dissolved (in the case of pulp) or melted down (for metals) or ground down (for plastics) and then reconstituted into new materials to make brand new containers. Sanitary and residual odor considerations will often preclude the fashioning of food packages from the recycled materials.

In the case of plastics, the recycling issue is further complicated by the range of materials known to the average consumer as plastics. (This is a very liberal assumption. The average person in many a meat packaging plant throughout the U.S. refers to the plastic packaging material as "the paper.") It is too much to expect of the householder to segregate plastics from other solid waste. To ask a further subdivision of the plastic discards would lead to total chaos. In 1988 the Society of the Plastics Industry suggested a voluntary coding system, identifying containers as to resin type. Inside a triangular "chasing arrows" logo is a number from 1 to 7 corresponding to PET (polyester), HDPE (high-density polyethylene), V (vinyl), LDPE (low-density polyethylene), PP (polypropylene), PS (polystyrene), and 7 for all others. The symbol identifying the resin is supposed to be placed on the inside of the container. The problems involved in getting the average person to segregate solid waste has been mentioned. It is an enormous chore to separate paper from the remaining garbage. Will the man in the street now be asked to identify and separate seven individual plastic types as well? In Italy, where the government has mandated the recycling of plastics, there is a plant that is shut down much of the time for lack of feed stock. There just is not enough plastic collected to supply the recycling process.

In the U.S. the recycling idea has not produced any significant results. It is possible that total elimination of the plastic component from the solid waste stream would not alter the massive size of the garbage mountain one bit. Plastics account for less than 8% of the solid waste tonnage. 92% of about 175 million tons of plastic-free waste is still more than local dumps can handle.

Plastics have been singled out by environmentalists because of their indestructibility (more under biodegradables below) and their low weight-to-volume ratio. On a density basis, plastics take up more room than paper, metals, or many other materials. Landfills are thus exhausted with fewer tons of plastics than with waste consisting of a mix devoid of polymeric materials. Especially plastic bottles seem to take up a very large volume and are therefore undesirable.

Some minor progress has been recorded with recycled plastics. A company in New York makes a plastic "super bag." The product is sold for about 50¢ in the supermarket as a reusable shopping bag. This bag, with a square bottom, can be reused about 10 times and then exchanged for a new one. The discarded bag is returned to the manufacturer for reprocessing. Similarly, a New Jersey manufacturer recycles PET bottles into scouring pads. Finally, some progress has been made with utilization of regrind in plastic bottle production. Plastic bottles are ground into a powder and this powdered resin is utilized for the center of a three-ply coextrusion, in which the inside and outside are made of virgin resin and the intermediate layer consists of the regrind. Other applications are still on the drawing board.

BIODEGRADABLE

One of the complaints heard against plastics is that "It's too good!" Most materials will disintegrate. They will be attacked by rain, oxygen, microbes, rodents, and other naturally occurring predators. Plastics will just lie there and not deteriorate. What is worse, plastic garbage bags will prevent their contents from decomposing. Thus the material shrinkage built into the landfill system has suddenly ceased to operate—exacerbating the volume problem even further.

One of the properties to recommend plastics was their durability and sealability. The food wrapped in paper would spill easily and could not be kept in the package for more than a few hours. The hermetically sealed plastic package could contain and protect its content for months or even years. Now this advantage turns into a liability. The ideal package is one that can offer the assured protection, but will self-destruct upon the invocation of some trigger mechanism. The choice of trigger is one of the pitfalls of this idea. Time is not a desirable cause for decomposition, since any delay in the distribution cycle may cause the package to disintegrate on the retailer's shelf. Light or temperature is not a suitable trigger either. It is very unlikely that the package will experience sunlight exposure during its useful life. Some plastics were designed to fall apart upon exposure to sunlight and were erroneously called biodegradable. They are not truly biodegradable, since the action is not related to microbial attack but rather on the chemistry of the plastic. What is worse, they do not degrade in the landfill. Yes, individual bags placed on a lawn and examined periodically will indeed disintegrate. However, they need weeks or months of sunlight exposure to promote this process. In the landfill, the plastic is dumped and has at most a few hours of direct sunlight exposure before the next layer of waste covers it and prevents any further sunlight from reaching it. The idea just does not work.

What is wanted is a chemical zipper which would keep the plastic tightly locked, but which could open on command. A tall order, indeed!

The cost of pollution abatement is rising sharply. In 1975 we spent 30 billion dollars, in 1990 it is estimated to exceed one hundred billion dollars. The price of cleaning up the environment is staggering!

AIR AND WATER POLLUTION

Packaging's impact on the solid waste problem is most visible. Nevertheless, the industry has contributed to air and water pollution, as well. Plastics are derived from petroleum products and must thus share in the pollution blame associated with oil exploration, transportation, and refining. Lamination, printing, and other processes, too, have spewed tons of VOCs (volatile organic compounds) into the air and washed millions of gallons of pollutants into the streams and rivers. Major success in controlling this type of environmental pollution have been reported in the past decade. Emissions have been drastically reduced through solvent recovery, afterburners, solvent elimination, and like procedures.

TOXIC WASTE

The cleanup of toxic waste dumps will take years and cost billions of dollars. The packaging industry is charged with complicity in creating these sources of poison storage. In fairness, it must be remembered that in most cases materials were disposed of in a legal fashion, consistent with the best scientific procedures of the time. If we were to apply the same logic of extended liability to other matters, we would experience a rash of lawsuits for failure to anticipate scientific discoveries of the last few decades.

CONCLUSION

The packaging industry is not blameless with regard to environmental pollution. However, it is not one of the primary polluters. The highways are not strewn with garbage dumped by food packers but by *people* who do not care enough to place their waste in designated recepticals.

CASE STUDY 16-1

One of the earliest recycling applications involved a carbon bed solvent recovery system in Canada. The new installation was 102% efficient (it was assumed that solvents from inks, which were not calculated in the source volume, increased the total recovered). The owners of this system were understandably satisfied. They had made an appreciable contribution to clean air and reaped the monetary rewards of practically eliminating solvent cost. However, after a few months the efficiency of recovery dropped and never returned to more than 65%. Large sums of money were expended in attempting to improve the system. But all efforts were to no avail. It must be assumed that the system was poisoned and eventually operated at about 35% efficiency. Not only was this no longer economical—but 65% of the solvent went up the stack into the atmosphere. Not a very good pollution control system.

CASE STUDY 16-2

McDonald's, the fast food chain, in late 1990 announced a discontinuance of their famous polystyrene package and will replace same with paper and/ or paper products. Environmentalists have scored a dubious victory. Under pressure, McDonald's capitulated and eliminated a functional package which will be nearly impossible to replace.

The one and only shortcoming of the polystyrene package was its large volume-to-weight ratio. Thus these trays contributed to the space taken up by refuse. This problem could have been minimized through compacting or recycling. The foamed polystyrene is easily identified and therefore more readily recycled than any other plastic material. The foam is suitable for regrind and could be utilized in many ways.

McDonald's may yet regret this change. Customers may find the paper wrapped meals to be cold and messy. Yet the environmentalists will not be pacified by paper wrap. After all, tens of thousand of trees will have to be killed to provide this new package.

ADDITIONAL READING

Anon. 1988. Plastics recycling: resin ID system is set; coded bottles soon. *Packaging Digest* **25**(5): 110.

Glenn, J. 1987. Do people tell the truth about recycling? *Biocycle* **28**(9): 56–57.

Goldberg, D. 1988. Charting progress:an update on mandatory recycling in the U.S. *Recycling Today* **26**(4): 42–45, 104–109.

Greenberg, E. F. 1990. Chasing arrows chase plastic recycling uniformity. *Packaging Digest* **27**(6): 30–33.

Kampouris, E. M. et al. 1987. A model recovery process for scrap polystyrene foam by means of solvent systems. *Conservation and Recycling* **10**(4): 315–319.

Keeler, J. H. 1988. Recycling plastics: an opportunity. *Journal of Packaging Technology* **2**(1): 4–6.

Moore, G. 1990. The potential of recycled plastics as feed stock for the production of melt-blown nonwoven webs. *TAPPI Journal* **73**(5): 93–95.

Nazarenko, L. 1988. New materials: polyethylene film is biodegradable. *Modern Plastics* **65**(4): 138.

Rickmers-Skislak, T. 1987. How to sell recycling programs. *BioCycle* **28**(9): 54–55.

Smoluk, G. R. 1988. PET reclaim business picks up new momentum. *Modern Plastics* **65**(2): 87–91.

Thayer, A. 1989. Limited impact seen for degradable packaging. *Chemical and Engineering News* **67**(10/23): 11.

Zucker, R. A. 1989. More waste-crisis issues discussed at conference. *Paper, Film and Foil Converter* **63**(6): 69–70.

Q 17: Which Is the Lowest Cost Package?

A 17: This question unfortunately has no simple answer. The question implies a concern for packaging cost as well as quality. To sacrifice protection of a costly product for the sake of economics in packaging seems ill advised. Thus one seeks minimal expenditure consistent with quality requirements.

It has been proposed that in the selection of packaging materials where cost is often a determining factor, several units of measurement may be considered. Many packaging materials are purchased and paid for on a per pound basis. Since the yield factor (square inches per pound or square meter per kilogram) may determine the number of packages that can be prepared from a given weight of packaging material, it too must be considered. The concept of cost per unit area is universally accepted and packaging experts employ this mode of cost comparison.

Price changes have at times been both rapid and substantial. Thus cost expressed on a per pound or per unit area basis, as shown in Table 17–1, require occasional revision. A nomograph, such as presented in Figure 17–1, could eliminate the need for updating price tables. For example, a plastic film with a yield factor 20,000 SI per pound was priced at 60¢ per pound. The new price just announced is $1.00 per pound. What was and will be its cost per thousand square inches? Drawing a straight line until it intersects the ¢/MSI column, it is found that the material of the example formerly cost 3¢/MSI and will now be priced at 5¢/MSI. Of course, for more sophisticated technologists the computer can handle this calculation (see Q 20 for a discussion of many computer applications in the packaging industry).

Furthermore, the cost per unit area could be compared on an equal thick-

Table 17-1.   Price Fluctuation of Select Packaging Materials.

		$/MSI ($/lb)	
	MSI/lb	1975	1990
Polyethylene (2 mil)	10.0	0.0360(0.54)	0.0467(0.70)
Ionomer (2 mil)	14.7	0.0700(1.03)	0.1308(1.50)
195 K Cello	19.5	0.0507(0.986)	0.1308(2.55)
Nylon (1 mil, coated)	21.2	0.0783(1.65)	0.1382(2.93)

MSI = one thousand square inches.

ness basis. Table 17-2 presents cost figures arranged according to increasing cost per pound. It will be noted that the ranking according to price per MSI does not at all correlate with that of the first column. Nor does the attempt to reduce all MSI prices to equal gauge restore the old order. Much less frequently utilized and often misapplied is the concept of cost per unit property. When a package is required to exhibit some specific property, cost per MSI is as useless as cost per pound in making the optimum material selection. What is most helpful under these circumstances is the cost per unit property comparison.

THE COMMON DENOMINATOR FALLACY

To compare the performance of various packaging material components it is necessary to find a common denominator for the measurement of such performance. For example, if one wanted to express the cost required to obtain a certain tensile strength, it would be necessary to convert the cost per pound (or cost per MSI) into dollars per psi. There are, however, two hidden fallacies in this approach which frequently escape notice and may lead to erroneous conclusion in the hands of a novice.

It is general practice to express tensile strength as pounds per square inch (psi) for a one-inch-wide strip one mil thick. When running actual tests on heavier or lighter gauges, the results are calculated to one mil. However, since in packaging certain gauges are more frequently employed, conversion to one mil does not agree with reality. If, for example, one faces the choice of using 50 gauge polyester or one mil Nylon, then a comparison of these two materials both calculated to the one mil thickness is not applicable to the problem at hand. This practice must be eradicated to lend meaning to practical test evaluations.

The second fallacy can be traced to the erroneous assumption that a common denominator can be realistically established. If the specification calls for a 1,000 pound tensile (per one inch width), one could make the selection

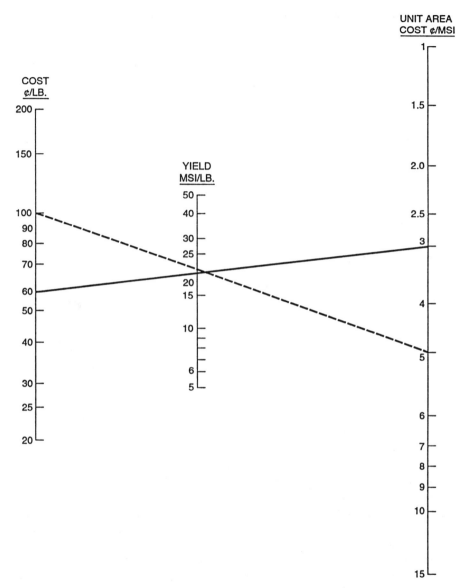

Figure 17–1. Cost nomograph.

Table 17-2.   Cost of Packaging Films.

	$/lb	$/MSI	$/MSI/mil
Polyethylene (LD 2 mil)	0.70	0.0467	0.0233
Ionomer (2 mil)	1.50	0.1308	0.1308
OPP (70 ga uncoated)	1.75	0.0438	0.0625
PET (50 ga uncoated)	2.20	0.0530	0.1060
OPP (70 ga coated)	2.26	0.0514	0.0734
105 K Cello	2.55	0.1308	0.1308
Nylon (1 mil coated)	2.93	0.1382	0.1382
PET (60 ga coated)	3.50	0.1042	0.1736
Aclar (22 A, 1 mil)	24.00	1.846	1.846

MSI = one thousand square inches.
1 mil = 0.001".

from Table 17-3 on the basis of cost per unit property. Column 4 of that Table seems to indicate that OPP is by far the least expensive plastic film (from the choices offered in Table 17-3) to accomplish the desired objective. In the real world this is not as clear cut as the theoretical consideration of Table 17-3 would make it seem. There is no 5 gauge polyester or 1.6 gauge OPP film available to achieve the desired low tensile strength. The films implied by Table 17-3, Column 4 just do not exist. The cost per unit property in this case is purely fictitious. An experienced packaging technologist looking at Table 17-3 would conclude that it would be no costlier and possibly slightly cheaper to use 50 gauge OPP in place of 25 gauge PET to achieve similar or even better tensile performance. The uninitiated, however, would be mislead by Table 17-3 in expecting huge savings from the substitution of OPP for PET.

THE ADDITIVE PROPERTY FALLACY

The concept of unit property cost is often inoperative because the assumption of additive properties is in many instances inapplicable. The unit prop-

Table 17-3.   Select Cost per Unit Property.

	Gauge (in.)	Ult. Tensile lbs × 1000	¢/MSI	¢/M lbs tensile
Polyester	0.0005	10	5.30	0.530
Nylon	0.001	11	13.82	1.256
Ionomer	0.002	9	13.08	1.453
OPP	0.0007	15	4.38	0.292

MSI = one thousand square inches.

erty costing is based on the assumption that property and gauge are in some way related.

While this is true in most instances, there are enough exceptions to make the additive property principle unreliable. Furthermore, modern packages are composed of sophisticated multilayer materials. No single material seems capable of meeting the many performance requirements. The properties of individual layers are not necessarily additive. Thus, for example, in a polyester–polyethylene composite the ultimate tensile is determined by the tensile strength of the polyester. The polyethylene has no measurable effect on the tensile performance of the composite. On the other hand, moisture vapor transmission rate is a function of the sum of the rate attributable to the components of the packaging material.

The principle of additive properties applied to the unit property cost concept can lead to absurd conclusions. Polyethylene is certainly cheaper than Aclar on any scale—including that of unit property. It would thus make economic sense to select poly rather than Aclar to obtain good MVTR. On a unit area basis poly is favored by a 50 : 1 ratio. On a unit property scale poly is still the favorite by a 2 : 1 margin. However, in practical terms one would have to replace Aclar with 30 mil poly—and package appearance, machinery problems, and other considerations eliminate this economical alternate as an absurdity.

SYNERGISM

While admittedly rare, it does occasionally happen that the total package performance is better than the sum of the individual components. It takes an expert to recognize this fact and utilize it effectively. The concept of unit property cost can be a valuable tool if knowledgeably applied. If misused it may lead to erroneous economic as well as package performance expectations.

Package cost evaluation should not be limited to a price comparison of various materials. The entire packaging operation must be considered in a determination of unit cost. One must assess labor productivity, shipping charges, the cost of returns, and many peripheral factors to arrive at a true unit packaging cost.

CASE STUDY 17–1

Years ago, DuPont circulated a study which compared the packaging cost of bacon, utilizing low-priced polyethylene in contrast to a more expensive ionomer composite. It turned out that the costlier packaging material actually saved the packer money by reducing in-house repackaging requirements and drastically reducing the returns from the field.

Sample Worksheet 1. In-Plant Leaker Cost Analysis

Polyethylene

Fixed overhead and equipment cost per minute	$1.02		
Labor—12 employees @ 9¢/man/minute	$1.08		
Packaging overhead and labor cost/minute		$2.10	
Packaging speed—packages/minute	70		
Packaging overhead and labor cost/package			
210:70-3¢		$0.03	
Cost of film, per package	$0.015		
Total packaging cost per package		$0.045	
In-plant leaker rate (%)	3		
Repackaging cost per million packages			
(0.045 × 0.03 × 1,000,000)			$1,350.00

Ionomer

Cost of film per package	$0.019	
Total package cost per package	$0.049	
In-plant leaker rate (%)	1.5	
Repackaging cost per million packages		
(0.049 × 0.015 × 1,000,000)		$735.00
Savings on in-plant leakers due to ionomer film		$615.00

Sample Worksheet 2

	Explanation	Amount
In-store leaker rate (polyethylene)	2%	
In-store leaker rate (ionomer)	1%	
Reduction in leaker rate	1%	
Total cost of package and meat	$0.50	
Gross savings per million packages		$5,000.00
Premium charges for ionomer per million packages		4,000.00
Net savings, per million packages		$1,000.00

Adding It Up

The total savings to be gained from reducing leakers, through the use of composite films based on ionomer film, is computed by adding the savings shown above. In the example shown, those savings per million packages are:

Savings on in-plant leakers	$ 615.00
Savings on in-store leakers	$1,000.00
Total savings per million packages	$1,615.00

CASE STUDY 17–2

Similarly, it has been shown that vacuum packaging cheddar cheese in a more costly packaging material actually results in very appreciable economics over the standard gas flush procedure.

1. The number of square inches of packaging material required for vacuum packaging is drastically reduced over those needed in the gas pack (Table 17–4).
2. Even though vacuum packaging materials are more costly, they are in the long run cheaper because of the lower area requirements. Thus, per pound of cheese the materials savings on an operation packaging 125,000 lb of longhorn may amount to $500/week.
3. The vacuum packaging equipment affords higher productivity. One vacuum packaging machine can package twice as much cheese as a gas machine. Labor savings can thus reach $200/week.
4. The gas flush package is covered by patent. Users of this mode of packaging may be subject to royalties of $\frac{1}{2}$¢/lb of cheese. Savings in royalty payments may amount to $625/week.
5. By switching to a more costly thermoformable material the cheese packer has a potential of saving $(500 + 200 + 625) \times 50 = \$66,250$ annually.

Table 17–4.   Material Requirements, Cheese Package.

	Square Inches per Package	
Package Size	Vacuum	Gas
8 oz.	44.23	67.5
12 oz.	47.87	80.75
16 oz.	55.18	90.11

ADDITIONAL READING

Ardito, G. J. 1973. Overspecifying wastes packaging dollars. *Canner Packer* **142**(9): 16.

Barrett, R. K. 1988. Talk about packaging convenience. *Canadian Packaging* **41**(2): 60–65.

Deanin, R. D. and Driscoll, S. B. 1973. Cost per unit property. *Modern Plastics* **50**(4): 110–112.

Flack, S. 1987. Packaging. *Forbes* **139**(1/12): 202–203.

Hirsch, A. 1976. Reexamine unit property cost. *Paper, Film and Foil Converter* **50**(8): 50–52.

Hunter, B. T. 1985. Packaging food. *Consumer Research Magazine.* **68**(2): 8–9.

Paris, E. 1984. Packaging (plastics upsurge). *Forbes* **133**(1/2): 224–226.

Vogt, H. V. 1973. Kostenoptimierung von verbundfolien und beruecksichtigung von gasdurchlassigkeit und steifigkeit. *Fette-Seifen-Anstrichmittel* **75**(5): 307–310.

Q 18: Does the Package
Improve the Product?

A 18: It should be remembered that all packaging involves compromises. Since there are many aspects to the selection of a package— it is almost impossible to achieve a perfect combination of properties. Our answer will be restricted to protection, display, and utility.

PROTECTION

The package may furthermore provide protection for its contents. There are of course several different levels of protection furnished. For example, good packaging assures the consumer measured quantity, unadulterated product, and hygienic handling.

Measured Quantity

The quantity of product purchased is stated on the package in either weight, liquid measure, or number count. The purchaser of 8 oz. of peanut butter in a sealed package is sure of receiving the stated quantity without the need for reweighing each package. Pilferage of a portion of the contents has been discouraged and is easily detectable (ex post facto).

Unadulterated Product

A good package protects the product from either accidental or willful additions or substitution of inferior products. The watering of milk, which was common in the early part of this century, has been totally eliminated by the introduction of sealed bottles and containers. There is of course no perfect guarantee against forgeries (often practiced with perfumes), imitations to mislead the consumer, and outright fraudulent substitutions.

Hygenic Handling

The package protects the product from dust and similar visible dirt. All of us have had occasion to pick up cans that have been sitting on the shelf for a while. The layer of dust on top of the can is amazing. Just imagine the amount of dust, insects, and all sorts of contaminants that would be present in product stored in open containers. In addition to the more obvious contaminants, there are microbes which are not readily visible to the naked eye. Transfer of bacteria may be affected by direct mishandling or through coughing or sneezing onto the product. A hermetically sealed package precludes such mishandling.

The package can maintain the product's quality through controlled atmosphere together with refrigeration or freezing. The package could even be pasteurized or sterilized and in this fashion afford increased shelf life for the product. However, the notion that the package can improve the product is definitely erroneous. One must remember the basic packaging principle— G-I-G-O (garbage in, garbage out). The most one can expect of a package is protection. The properly designed package can preserve the product, it cannot improve it.

Far too often the sins of the packer are blamed on the package. Product deterioration is in most instances irreversible. Once flowers are wilted no magic package on earth will make them bloom again. The best one can expect of a package is the preservation of the status quo.

There are of course many other factors which enter into the selection of the protective package. One of the most important is economics. While money talk is often obnoxious to the average packaging technologist, it is of vital concern to all the other business functions.

For example, the best protection for any food product is offered by a tin can. Is such protection really needed to package candy? Can the product afford the cost involved? Even if the answer to the preceding question were in the affirmative, there remains the question "Is it desirable?" Would the consumer want to bother with extracting the candy from a tin can?

While the example given may sound somewhat absurd (after all, who would propose to package candy in a tin can?) the sin of overpackaging beyond belief is not a rarity. The problem of which is the best package, even for the single objective of protection, remains unanswered. The example of the tin can was meant to illustrate the undesirability of overpackaging. However, even in the range of flexible materials overpackaging is a distinct possibility. In vacuum packaging one pound of bacon one normally employs composite films consisting of not less than 0.00075" and not more than 0.00125" Nylon with or without PVDC and a layer of not less than 0.0015" and not more than 0.003" polyethylene. In place of polyethylene one can substitute EVA or ionomer. The combinations and permutations that can be achieved with these variables (keeping gauges at $\frac{1}{4}$ mil intervals) is 126 distinct composites. Within this range one can find at the lower end of the spectrum packaging materials which are barely adequate and at the upper end an array of materials which are possibly slightly in excess of what is required. Further improvements in the quality of materials increase the cost of packaging without improving the package performance. To switch to 0.002" Nylon from any of the above satisfactory composites will not improve the packaging performance to any measurable degree.

To summarize, for protection's sake the best package is the one that will provide the maximum required protection at the minimum cost. Other factors must of course be considered, as well.

DISPLAY

The product often has consumer appeal and thus sells itself if properly displayed. There are many products that depend on demand appeal. Fresh meats, fruits and vegetables, cheeses, processed meats, and many other items seem to emphasize esthetic display. Some cookies and cakes take advantage of visual display to create impulse demand. In general, the transparent package can provide product display. This property can be beneficial if the product is attractive and can withstand exposure to light and the diminished protection offered by transparent materials.

The package can improve the visual appeal of some products. For example, potato chips are not very attractive. Some are burnt, others are crumpled. An opaque package with attractive graphics can not only hide the imperfections of the product but actually provide visual consumer appeal through the printed image. The same may be applicable to a large array of other products. Pet foods, frozen fruits and vegetables, cereals, cookies, and many others may be functional, nutritious, delicious, of good quality—but they are certainly not attractive looking. An opaque pack with imaginative pictorial display may enhance the product's acceptance. Soup mixes, which are neither beautiful to behold nor shelf stable within a transparent package are therefore sold in foil packets. The appeal of such packaging can be enhanced by surface graphics to improve consumer recognition and demand. The "best" package will consider both the esthetics of the product as well as its shelf stability in selecting the most appropriate packaging material.

UTILITY

The package in addition to its visual appeal and the protection it affords the product must prove useful. The can is a good example of a package with increased utility. The old standard can for beer and soft drinks was a good package. It did, however, require a can opener and another container to transfer the beverage prior to consumption. The new, improved, easy-open can eliminated the need for both of these accessories and is thus of greater utility to the consumer. The best package would provide maximum utility to the packer, the retailer, and the ultimate consumer.

CASE STUDY 18–1

A meat packer complained of his packaged frankfurters turning green. He accused the packaging material of being contaminated. Bacteriological examination of the flexible plastic packaging material proved negative. The suspicion that the packaging equipment failed to provide adequate vacuum proved equally without foundation. The search continued for imaginary culprits. Not once did the packer consider that his own lack of sanitation was at fault.

CASE STUDY 18–2

A nut packager complained regarding the off flavor developed in one lot of his packages. The residual oxygen levels were checked and found satisfactory. The barrier properties of his packaging material were found to be within specification. One must assume that the nuts were at least slightly rancid prior to packaging. Since the package was transparent, the autooxidative rancidity action continued, aided by sunlight. The substitution of opaque packaging material reduced the problem appreciably.

CASE STUDY 18–3

This is a well known case of the "best" package turning out to be a disaster. A supplier of dry soup mixes was convinced by the high-powered sales organization of a giant resin supplier to switch to their specialty sealant resin. Claims for that resin were "fantastic."

The packer did indeed switch his entire line of products to this miracle sealant. After a few months, the returns of certain items started to roll in. It seemed that most mixes held up well. However, all products containing chicken fat were plagued by delamination and seal failure. The miracle sealant had no tolerance for chicken fat.

The moral of this event—do not accept claims of suppliers blindly. Test the product in the proposed package prior to going to market.

ADDITIONAL READING

Anon. 1988. Comment: combating the anti-packaging lobby. *Packaging Technology and Science* **1**(1): 1-3.

Anon. 1977. Safety in eating due to packaging. *Good Packaging* **38**(9): 55.

Anon. 1977. Packaging—preserving and presenting new foods. *Food Processing* **38**(11): 46-50.

Anon. 1977. What users think of flexible packaging. *Packaging Digest* **14**(4): 20-21.

Anon. 1976. Flexibles: high barriers for fresh foods. *Modern Packaging* **49**(7): 21-24.

Anon. 1976. International conference on the protection of perishable goods through packaging. *Neue Verpackung* **29**(8): 798-801.

Irgel, L. 1977. Marketing: well packaged is half sold. *Verpackungs Rundschau* **28**(7): 932-934.

Pintauro, N. D. and Simah, A. H. 1976. Teamwork: your key to longer shelf life. *Package Engineering* **21**(4): 38-40.

Shaw, F. B. 1977. Toxicological considerations in the selection of flexible packaging materials for foodstuffs. *Journal of Food Protection* **40**(1): 65-68.

Q 19: What Is a Flexible Can?
How Can It Be Used?

A 19: All of us are familiar with canned food products. Anyone who has set foot in a supermarket has seen shelves filled with canned soups, vegetables, fruits, and many other items. Cans have proven themselves over several decades. They are a reliable means for preserving food items at ambient conditions for very prolonged storage cycles (several years, if needed). A can is a metallic container which is capable of maintaining a hermetic seal. Foodstuffs are placed therein and after the metallic can has been sealed it is subjected to a heat sterilization cycle. All bacteria associated with the product are killed in the retort operation, and thus the product in the sealed can may be maintained without spoilage for an almost indefinite period of time at a wide range of temperature and humidity conditions.

The advantages of the can are many. Even though the metallic container is costly, the packaging process is suitable for exceedingly high-speed operation (a thousand cans per minute or better) and thus becomes economically attractive. The packaged product is well protected under all sorts of adverse storage conditions. The sturdy storage container can withstand unusual abuse and thus minimizes losses due to shipping and handling.

The relative ease of accessibility must be considered a further advantage of this packaging mode. Entry into a can has been further improved in the early 1970s with the pull tab top. A further convenience feature available in most if not all canned foods is the degree of preparation. Most items are ready for consumption. Good examples are beer and soft drinks, canned fruits, and a large assortment of other items. Where some preparation is required, it has normally been minimized. Thus, for example, fresh carrots bought in a plastic bag must be peeled, diced, and cooked—while those bought in a metal can require only heating prior to serving. The canning company has taken on the chore of peeling, dicing, and partially or fully cooking the product.

Cans are by no means the ideal package. They do have a number of serious disadvantages. Cans are bulky and heavy and thus add shipping weight and storage space requirements. Because of the large volume of food contained within the can, the package has to be overheated during the sterilization cycle in order to assure the minimum required heat to reach the center of the package. Thus the food at or near the wall of the can gets overcooked and loses desirable flavor and texture. Very often the can itself imparts a characteristic flavor to the food product.

THE FLEXIBLE CAN

Historically the search for a replacement of the tin (or aluminum) can has been promoted by the armed forces. Especially the Army's Natick Laboratories have been searching for ways and means to provide the military with individual meals that can be carried in the soldier's pocket or knapsack. Cans (K or C rations) were prevalent during World War II. The army felt that the weight of the can, its bulkiness, and its potential hazards were all in its disfavor. In the early 1950s the army experimented with radiation sterilization of food.

Raw, semi-prepared, or fully prepared portions of food product could be placed in a flexible pouch and sterilized by exposure to cobalt or other sources of radiation. There were many problems associated with this process. However, this approach to food preservations was finally abandoned when the FDA refused to grant approval. In the early 1960s Natick Laboratories, in conjunction with Continential Can Co., Rexham, Swift & Co., and others, embarked upon a coordinated effort to produce a flexible can. In essence what the Army wanted was a foil-containing pouch which could go through a heat sterilization cycle and preserve food products for a prolonged room-temperature storage cycle. Some of the more specific problems facing the Army Quartermaster Corps will be taken up in subsequent paragraphs.

While the American effort was proceeding at a snail's pace the Japanese commercialized the flexible can on a large scale. On the European continent the flexible can made its appearance in the early 1970s, and in Canada in the mid-1970s. The FDA withheld approval of this package until early 1977. Primary concern dealt with the possible migration of low-molecular-weight polymers, adhesive components, plasticizers, etc. into the food product at elevated sterilization temperatures. Most of the materials employed in the flexible can do possess FDA approval at room temperature and/or for limited-time exposure at 212°F (100°C). For example, the boil-in-bag has an extended history of safe use in the U.S. However, elevated temperature exposure is limited to a few minutes and in most instances less than 212°F. The FDA insisted on migration studies involving possible decomposition of adhesives or plastics at 250°F and quantitative determination of migrants into the food, treating same as adulterants with appropriate feed studies.

Eventually, the FDA accepted such study conclusions and approved the flexible can. In the meantime, Continental Can and Reynolds Metal both marketed flexible can materials based on approaches that avoided facing up to the issue of extractables. With the advent of FDA approval there has been appreciable activity in the U.S. aimed at stimulating demand for this new packaging approach. The huge success of the flexible can in Japan was cited as proof of a potentially large market in the U.S. Proponents, however, failed to grasp the difference between the American and Japanese market which precluded any parallelism. Lax sanitary standards require all packaged food to be pasteurized in Japan. Since such a post-packaging step is absolutely required, the additional time and temperature needed to achieve full sterility is not of major concern. In the U.S., on the other hand, the overwhelming majority of flexible packaging does not go through any heating cycle and thus retort would increase the cost of an otherwise inexpensive packaging system. In Japan, the quantity of food in general and especially of meat consumed by the average resident is rather small. Thus a small package of 4 to 8 ozs. may serve the needs of the family. In the U.S., on the other hand, much larger portions are standard fare. A family-size retort pouch seems to negate the advantages the pouch has claimed over the can. Another important marketing consideration which may have been overlooked in comparing the Japanese and possible European consumer reaction to the retort pouch involves the ready availability of refrigerator and freezer space both in the retail establishment and the average home. On the foreign scene there may indeed be a need for shelf-stable food, simply because of limited refrigerator and freezer space. On the domestic scene, there exists no such need. When this reality was finally reluctantly accepted, there were attempts to address specialty markets (gourmet foods) or selected end uses (hunting and camping). In the end, no mass market demand ever developed.

THE NEED

The army has definitely established a need for a flexible can to feed the armed forces in the field. This new package will not replace normal feeding patterns on army bases nor even eliminate field kitchens where same can be set up under combat conditions. The retort pouch is merely a supplement to the normal feeding provisions to provide emergency rations for the soldier who will be unable to make use of standard feeding facilities.

There is a definite need for shelf-stable food items for leisure activities. As our working week dwindles down to four days, more and more people will be participating in activities such as fishing, camping, and other long-range outdoor recreation which will take participants away from normal feeding establishments. Availability of shelf-stable food items in lightweight, flexible packaging will provide easy preparable meals under such circumstances. The retort pouch would be an ideal conduit for distributing food to underdeveloped countries. Very frequently food sent to starving nations spoils prior to reaching the hungry, for lack of proper storage facilities. Unfortunately the poor cannot afford the food much less the expensive package.

The gourmet food market expects increased volume distribution via the retort pouch. As explained earlier, less heat exposure is required in the retort pouch and thus the product tastes more like freshly cooked rather than canned.

THE PROBLEMS

The flexible can is not without its own set of new problems. Some of the difficulties associated with the metal can have been eliminated only to have a complete new set of problems arise.

Speed

Canning operations have been automated over many decades to a point where lines filling more than a thousand cans a minute are the rule. At this writing the flexible can is limited to approximately 30 pouches per minute, and looking forward to increasing this to about a maximum of one hundred per minute sometime in the future. It would thus take 10 to 20 lines to replace one of the existing manufacturing facilities for switching from the metal can to the flexible one. To produce packages in large volume seems thus out of the question, and the flexible can is in essence restricted to the high-cost, limited-volume item. Even the next generation of the flexible can made from thermoformable foil composites on a flatbed machine would still face a productivity limit of under 60 packages per minute.

Evacuation

The flexible can should contain a very minimum of residual air. Most products packaged are oxygen sensitive (containing fats) and must thus be protected from turning rancid. Not only does this require good evacuation prior to sealing of the package, but it also mandates a super barrier material in order to exclude reentry of oxygen during storage. Thus all of the flexible cans are foil composites rather than transparent flexible packaging materials. This requirement for foil increased the packaging cost appreciably. The very same product sold frozen could be packaged in a poly bag at a small fraction of the packaging cost. Evacuation of the pouch is also necessary in order to reduce the internal pressure during the retort cycle. The pouches would have a tendency to explode as the temperature in the retort increases and the residual air within the package expands. Some of this ballooning of the retort pouch can be counteracted by retorting under water pressure. However, the best method of minimizing the internal pressure problem involves reducing residual air. Since most food items have either small particles (such as kernels of corn or grains, etc.) or gravy associated with them it is difficult to achieve a high degree of vacuum. If the contents are relatively mobile, such as liquids or dry grains, then portions of the contents may actually be transferred into the vacuum line. Not only does this create a mechanical problem but also one of maintaining proper content of the pouches. Currently the preferred process of evacuation is one of injecting steam into the pouch prior to sealing. As the steam cools and condenses a partial vacuum is created.

QUALITY CONTROL

Billions of cans have been produced, filled, sealed, shipped, sold, and utilized by the consumer. A high degree of reliability has been established which requires minimum QC assurance levels. No such reliability has at this date been comparably established for the retort pouch.

What is more disturbing is the fact that no reliable nondestructive QC test exists for evaluating the performance of the flexible can. Natick Laboratories spent many years in developing test procedures for heat seals. To this date there is no reliable, foolproof nondestructive heat seal test. The same would apply to pinholes of a microscopic size or a variety of other nonapparent potential failure mechanisms. A 100% inspection of all pouches is not only costly and time consuming but also, because of the destructive nature of most tests, absolutely impossible. It is hoped that in several years enough historical data will have been collected to provide a basis for statistical evaluation of batch processing.

PACKAGE SURVIVAL

The retort pouch is subject to rips, tears, and punctures. The flexible material is relatively strong—as far as flexible materials go. However, it in no way compares to the strength of a metal can. In order to protect the pouch from inadvertent damage, Natick Laboratories and its subcontractors have elected to enclose the flexible can in a corrugated sleeve. This mode of protection adds to the bulk and the weight and thus countermanding the original objectives of the flexible can. The board sleeve furthermore adds cost and reduces the productivity, making the flexible can even less competitive with the metal container.

THE FLEXIBLE CAN TODAY

Two composites are currently in use in the U.S.:

Polyester/aluminum foil/cast polypropylene
Polyester/aluminum foil/modified high-density polyethylene.

These composites utilize urethane adhesives or extrusion lamination. The size of the potential market is still in question. No doubt the army will continue to be the primary outlet for such packaging. However, considering the reduced size of the U.S. Army and the fact that at best a small portion of its feeding requirements will be in retort pouches, the potential size of this market, too, is rather limited. The future of the retort pouch will continue to be nebulous.

CASE STUDY 19–1

A major canner of pineapples decided to follow the trend away from the can and offer the product in a novel package. After considerable deliberation, the new mode selected was fresh frozen. Chunks of pineapple were placed in a transparent, thermoformed plastic package and quick frozen. The package weight and cost was reduced appreciably. Nevertheless, it did not sell.

Consumer interviews revealed many reasons for the reluctance to embrace this "improved" product. The dominant complaint was "it doesn't taste right!" For decades the average American eating canned pineapple had been trained to expect a certain taste (slightly cooked from retort and tin reacted with the acid juice). Now, the new product lacked tin and was really fresh. It just did not taste like pineapple.

An improvement is no improvement if nobody wants it!

ADDITIONAL READING

Anon. 1987. Retortable and microwaveable plastic pouch. *Food Engineering* **59**(2): 48.

Anon. 1977. Special today: the retort pouch. *Packaging Digest* **14**(9): 34–41.

Anon. 1975. Economic pressures bring temporary halt to the retort pouch growth. *Packaging Review* (UK) **95**(10): 83–89.

Andres, C. 1977. Thermally processed shelf-stable foods have quality equal to fresh prepared. *Food Processing* **38**(9): 30.

Freshour, O. 1987. Time for retort pouch may finally have come. *The National Provisioner* **196**(12): 14–19.

Garland, T. D. 1976. The retort pouch—a report of importance from Swan valley. *Canadian Packaging* **29**(3): 26–29.

Kelley, N. 1988. High barrier nonmetallic retortable packaging material. *Activities Report of the R & D Association* **40**(1): 46–47.

Kramer, A. 1976. The retortable pouch poses no threat to frozen foods. *Quick Frozen Foods* **39**(4): 244–246.

Martin, S. 1975. A retort to the retortables. *Quick Frozen Foods* **37**(11): 16.

Meeker, G. 1988. Atomic edibles? *Health* **20**(1): 65–68.

Mock, D. 1985. Star wars comes to the supermarket. *Science* **6**(9): 76.

Rice, J. 1988. Retort plastic packaging: an idea whose time has come. *Food Processing* **49**(3): 70–78.

Weber, L. 1988. Food irradiation update. *Good Housekeeping* **207**(7): 205.

Q 20: How Are Computers Impacting Packaging?

A 20: The computer age has overtaken the packaging industry. Those who failed to recognize the many facets of computer control and utilize them to advantage have been left behind to suffer the fate of the quill pen. The modern business relies on computer assistance in all its operations. The computer has offered these advantages:

  a. Faster decision making.
  b. Speedier problem solving.
  c. Quality improvement.
  d. Higher yields.
  e. Bottleneck elimination.
  f. Inventory reduction.

The quality, quantity, and timeliness of information is improved, thus permitting managers to make intelligent decisions quickly. Computerization can result in improved product quality, accountability, and profit. It can also result in better production machinery monitoring, SPC (statistical product control) and comprehensive reporting.

BUSINESS FUNCTIONS

The computer has invaded the normal business functions in the packaging industry just as in any other commercial concern. Order entry, accounts receivable and payable, inventory, personnel, etc. are all controlled by computer information systems. In a well integrated system these functions are not just independently handled but are linked, so that a new order automatically triggers several responses. Purchasing is alerted to order raw materials. Scheduling places the new order into the production queue. Customer service notifies the customer of anticipated delivery date. And on and on—the computer keeps track of the order until it is shipped, billed, and paid. And even then the order is not forgotten. The information is stored and provides sales/marketing a history on the customer's order pattern and on product sales volume.

The computer is thus a valuable tool, supplying a wide range of up-to-date business data, enabling timely intelligent business decisions.

PACKAGING MATERIALS

The combinations and permutations of materials available for package production are legion. A trial-and-error approach to selecting the most suitable and yet least expensive material for a specific application would be very costly and time consuming. More about this under the topic of Packaging—Design (CAD). The converter is assigned the task of producing the material from which the package is made. The processes involved may include printing, coating, laminating, extruding, slitting and others, all of which can involve computer control.

Most computer systems are custom designed for a specific application. Often a system that works well for one converter cannot be transferred to a competitor without some major modifications. It is therefore not prudent to recommend any hardware or software as being favored for a given situation. When names are mentioned, they are offered as sources of initial contact, only. The reader is advised to seek other suppliers and/or expert help for the specific problem at hand.

Printing

There are several printing processes utilized in the packaging industry. The two most widely employed are flexographic and rotogravure roll stock printing. The problems most often encountered are color variation and misregister. The process operates at very high speed, which makes it almost impossible to check these parameters. Stopping the press in order to visually inspect the printed material reduces productivity and offers no guarantee of compliance with specification. The sample for review is run at much lower than operating speed and may indeed comply, but all the material run at 1000 fpm may be defective. CCI, Inc. (Hackettstown, NJ 07840) claims to have an automatic detection system. It provides programmed scanning for continuous or selective viewing of each image. The computer can react to the information furnished by making adjustments to the ink viscosity to alter the laydown and in this manner the color depth. By changing web tension or speed the register can be adjusted as well. Defective impressions can be flagged for future removal.

Lamination

There are even more process variables in lamination than in printing. Ideally all should be carefully controlled. Practically, it may be too costly and time consuming to deal with any but the most critical ones.

A laminator builder, such as Faustel, Germantown, WI 53022 can be of assistance. The system selected could be very simple or very complex, depending on the functional requirements expected of the control system. For example:

a. How many I/O capabilities are desired? On a laminator the amount of input information may include: speed, unwind tension, rewind tension, adhesive viscosity, adhesive laydown, skips, oven temperature, oven air velocity, solvent concentration, nip temperature, nip pressure, side-to-side alignment—and in some instances also level of corona treatment, face-to-face print alignment, solvent residue in laminate, opacity, static charges, etc.

b. Closed loop process control. The data acquired could be made available to the operator only. He in turn would need to react and make needed adjustments. As long as equipment is operated at low speed (50 fpm), a few flashing lights or bells can alert a responsive employee to the fact that certain parameters have drifted outside approved limits. The operator would chase around the machine, making adjustments and paying little attention to the overall operation. At today's speed, this type of operation would be virtually impossible. The data gathered from the various functions must call for automatic adjustments to keep the operation on keel.

c. Information sharing. Does the system—or should it—share its information with other departments? For example, is the data generated made available to QC? This information in the hands of QC may eventually yield statistical data to improve the process and consequently the product. Another instance of the importance of information sharing: the laminator had some mechanical problems and after spoiling ten thousand feet of material, the scheduled run was aborted. Purchasing would need to order replacement material, customer service should notify the customer to anticipate a delay in delivery. Unless the processing information is shared with all the other departments, the system is liable to prove of limited value.

d. Information receipt. Processing must receive information from other departments in order to function efficiently. This information exchange must be tied in with the processing computer. For example, if a change has been made in one of the raw materials then the processing computer must be informed, or else the new material will be rejected as unsuitable.

PACKAGING

Similar considerations apply to the packaging operations. Computers are indispensable in today's business environment.

CAD (Computer Aided Design)

The designing of a new package was a lengthy and very costly process. It took months and even years to bring a new package to market. Each minute little change required a complete redrawing of the package design. However, with CAD the process is drastically reduced both in time and tedium. For example, ten shades of the same color can be evaluated within a few minutes and samples of each produced in the same time frame. This very same end result would have required days of an industrial artist just a few years ago.

Quite a few CAD programs are available; each has features which are preferred by some and disliked by others. Beatrice/Hunt Wesson claims that the VASTER design and imaging system from DuPont Design Technologies has enabled them to complete a new package design in two months. Purup NA, St Paul, MN 55108 is claimed to handle software programs such as Aldus Pagemaker®, Quark Xpress®, Aldobe Illustrator®, Ready Set Go®, Ventura Publisher®, Imagestudio® and others.

Bar code, too can be designed and evaluated by computer technology. Software for GBL (generalized bar code and labeling system) is available from several sources, such as Integrated Software Design, Mansfield, MA 02048.

CAM (Computer Aided Manufacture)

At the packaging machine, the integrated computer system is an effective means to set up, monitor, and control the numerous process variables. The operational benefits to be derived include:

a. Display of key machine variables (with hi/lo alarms).
b. Storage and display of process condition specifications (possibly with alarm limits).
c. Continuous monitoring of machine variables (such as speed, seal temperature, print register, etc.).
d. Recording of machine variables. This could be for critical parameters

only. It could be continuous, or at set intervals, or only to indicate unacceptable intervals.

e. Real-time statistical process control (charting select variables).
f. Job identification (yielding a job summary report).
g. Shift summary report.
h. Preventive maintenance of critical components.
i. Many others.

The computer's informational environment consists of:

Network—the communication path between the computer and devices it helps to operate.
Input—the information available.
Output—the relay devices available.

The network is divided into levels, each having a separate function. The lowest level is the machine level. Information is communicated from the machine to the sensors and the programmable logic control (PLC). Information from the PLC to the actuators represents an output that causes the actuators to make an adjustment in the machine's operation.

The next higher level above the machine level is the station level (SLC); it involves a microcomputer. Here information passes between the SLC and the PLC. It has access to all the input and output data of the PLC. The SLC provides the operator with information regarding the current ongoing operation.

The next highest level is the cell level. This is usually not at the manufacturing station, but at QC or manufacturing control office. It functions as an off-line computer and is mainly involved in the collection and analysis of historical data from the SLC. This computer is normally more sophisticated and can perform several operations simultaneously, thus interacting with more than one SLC.

Interested parties may want to contact the Packaging Machinery Manufacturers Institute, Washington, DC 20005, for information regarding packaging machinery in general and computer controls specifically. Technology Automation of St. Paul, MN 55164, claims to have several packaging line installations to its credit.

Quality Control

The health of a manufacturing business depends to a large extent on the quality of its products. The QC department can and should be linked to manufacturing through the computer system and thus develop statistical product control (SPC). The ULTRA CHECK SYSTEM® by KRONES (414)

421-5650, claims to be able to acquire, store and process data for SPC. Many other such systems are available. The QC department will want to run independent tests on random samples or retains. Each tester can be hooked in with computer control to plot test results, obtain averages and compare the new results to specification requirements and historical data on the same product. Tests on retains can help to develop an aging history of a product.

Finally, there are many math and graphic software packages available to assist in the preparation of graphs or the resolution of mathematical problems. TECH*GRAPH*PAD from Binary Engineering Software, Inc. (Waltham, MA 02154) might be of interest in this respect. *Chem-Calc,* distributed by the American Chemical Society, Washington, DC, is a program designed for chemist's calculations, while *Polymat PC* is a database of numerical properties and text data on thermo-plastics, etc. The database is maintained by German Plastics Institute.

There are many more computer applications in practically every business department. Some are specific to just one phase of the business. Nevertheless, permitting others to share the information may have beneficial results for all. For example, STN Express (International) affords access to databases such as CAS ON LINE, Chemical Journals online, COMPENDEX, INSPEC, and others. For the R&D department this broad range of information is indispensable. For purchasing, manufacturing, or accounting this information is too sophisticated to be of much value. However, process engineering might find data impacting on the local process and thus yield better quality, improved productivity, and/or lower processing cost. The example offered places the technical data at the disposal of a technical trained employee of an other discipline. Yet the company can benefit even by providing this same information to technical novices in the marketing department. An alert marketing individual may find in a review of the technical literature the seeds for a new product or a new application for an existing one.

CASE STUDY 20-1

Many converters, both large and small, have in the last 20 years faced bankruptcy, takeover, and shutdown. There are many reasons for this state of affairs. However, in one particular case the cause was very evident.

The general manager of a medium sized converter had over a period of 30 years advanced from messenger to the top position. He remembered and often mentioned with pride that he had introduced the first adding machine into the firm. This was his one and only leap forward and he remained on this plateau for the rest of his commercial life. While others introduced desktop computers for order entry, personnel records, etc., our regressive manager stuck with pencil entries. In 1988 his office had no word processors and his laminators ran at 50 fpm without computer controls. The cost of his products spiraled upward, while their quality decreased constantly. Is it any wonder that this enterprise has expired? A computer-free business has no business!!!

ADDITIONAL READING

Anon. 1990. New packaging design created in 2 months. *Packaging* **35**(13): 117.

Anon. 1986. Computers ease path for barrier containers. *Modern Plastics* **63**(3): 30.

Anon. 1985. Computerized shelf life system tracks and grades product freshness. *Food Engineering* **57**(8): 125.

Brody, A. L. 1986. Compupac '86 provided a preview of computer design. *Food Engineering* **58**(8): 54-56.

Hartley, J. J. 1990. How do computers relate to coaters/laminators? *Paper, Film and Foil Converter* **64**(4): 60-61.

Hever, R. 1986. Computerized checkweigher in—human error out. *Packaging* **31**(12): 77.

Hushon, J. M. 1990. Expert Systems for Environmental Applications. Washington, DC: American Chemical Society Symposium Series No 431.

Klinetob, R. 1990. Technology for competing in the 1990s: Millwide information management. *TAPPI Journal* **73**(6): 261-265.

Larsen, M. et al. 1990. The packaging line revolution. *Packaging* **35**(10): 33-34.

Marsh, K. S. and Wagner, J. 1985. Computer model looks at the environment to predict shelf life. *Food Engineering* **57**(8): 58.

McLellan, M. R. 1985. An introduction to computer based process control in a food engineering course. *Food Technology* **39**(4): 96-97.

Ouchi, G. 1986. *Personal Computers for Scientists: A Byte at a Time.* Washington, DC: American Chemical Society.

Puhlick, A. F., Jr. 1990. Using a computer in a pressure sensitive adhesive line. *TAPPI Journal* **73**(6): 97-100.

Witt, C. E. 1985. Computer assisted packaging offers more options and accuracy. *Material Handling Engineering* **40**(9): 109-110.

Q 21: What Problems Does One Encounter with the Packaging of Frozen Foods?

A 21: In addition to all the problems present in ordinary food packaging, there are several exclusively related to frozen food. Low temperature (about 10–20°F) retards bacterial degradation. Under these circumstances the packaging material is not called upon to furnish the type of protection required for refrigerated foods. The primary function of the package is thus relegated to "contain" and "display." The package may be as simple as a polyethylene bag or a waxed box. These materials provide the primary protection required—moisture barrier. The atmosphere in the freezer case is relatively dry. Products placed in this environment are in danger of drying out—freezer burn. Packaging materials with good MVTR are thus required to overcome the moisture loss tendency. Polyolefins and waxes are prime contenders for this service.

The selection of the packaging material must also consider the severe abuse conditions resulting from freezer storage and display. Most plastics suffer diminished impact or flex-crack performance at lower temperature. A choice of polystyrene would be most inappropriate. This plastic material would perform worse than thin glass.

One must further consider the changed physical characteristics of the packaged product brought about by the freezing process. It is, for example, relatively easy to contain diced carrots. Whether fresh or partially cooked, the carrots are not prone to cut or puncture the packaging material. However, when frozen the otherwise soft edges turn into knives, the corners into spears.

The package must take a high degree of abuse under these conditions. Frequently, the frozen product is thawed or heated in a plastic package. Consideration must be given to this additional process, as well. What are the directions and what is the consumer likely to do? The directions may spell out clearly that the package of frozen vegetables is to be placed in a pot of boiling water and kept therein (no cover on the pot) for 15 minutes. We know full well that some housewives will improve on these instructions by placing the cover on the pot and thus raise the boiling temperature by several degrees. Another consumer will forget the pot on the stove for up to an hour. In either case the package should survive the abuse.

Many frozen products are packaged with sauces. These fluids are automatically injected into each package. Frequently, however, drippings from nozzles contaminate seal areas. Such contamination, even though wiped clean, may lead to weaker seals. The proper selection of sealant can minimize the problem considerably. Of course, the mechanical elimination of dripping is the best means of avoiding the problem.

Two developments of special interest should be mentioned here. *Fresh meat* sold in the frozen state has found some limited acceptance lately. The BIVAC system (see Q 28) thermoforms an unconverted ionomer film around the meat portion. Ionomer provides adequate moisture barrier to preclude freezer burn. However, this film has a high oxygen transmission rate, which maintains the bloom of beef cuts during freezer storage. The package seems to have found wide application in the institutional market. The other frozen food development of recent vintage deals with prepared meals. TV dinners or a large assortment of individual courses were sold in aluminum trays, ready to heat and serve. Many objections have been raised to the metal tray:

- Cost.
- Unsuitable for microwave heating.
- Hazardous to handle (burns and cuts).
- Disposability—will not incinerate.
- Serving—requires other dishes to serve and eat.

Recently, attempts to overcome these objections have taken two approaches. Plastic-coated board has been introduced especially for bakery items. There are also experiments underway to utilize trays made from polyester-coated board for microwave reheatable luncheon service. At the same time there is some activity in the market with regard to plastic packages that can overcome all of the aforementioned objections. Several plastics have been mentioned as meeting all the requirements of microwave as well as conventional oven heating. Polysulfone, polymethylpentene, and polyester have all been considered for this application.

BOIL-IN-BAG

The introduction of frozen food naturally brought about the adoption of the convenience feature of the boil-in-bag. This package provides for heating the product in the sealed package. It permits the consumer to prepare the food in a readily available pot or pan with hot water. Since the product never soils the pot there is no need for clean up.

The technology of boil-in-bag is a perfect example of the confused state of the art pervading this industry. Polyester/medium-density polyethylene was found to perform well and was accepted universally as the packaging material for this application. Specifications were written based upon the actual performance parameters of this composite. One of the physical constants called for a seal strength of 10 lb/1″ width. This requirement has survived for many years. Newer materials have been introduced to the frozen food industry which have survived many hours of boiling tests. These materials could have saved the industry thousands of dollars. However, the new materials failed to provide the sacred 10 lb/1″ of seal strength and were consequently rejected. The fact that they performed satisfactorily under actual use conditions was of no consequence.

A frozen food packer raised an important objection to the above named test. Most tests, including that of seal strength, are conducted at room temperature. The 10 lb seal has proven itself most reliable at room temperature. A chamber was constructed to test seals at elevated temperature and/or while submerged in any liquid. This modified test revealed that the seal strength dropped off rapidly at elevated temperature.

This newly acquired insight caused many packers to switch to polyester/modified high-density polyethylene. This composite, too, suffers from diminished seal strength at boil. However, the residual seal strength is usually in excess of 2.5 lb/1″ width, which is more than adequate to maintain the integrity of the package.

In summary the packaging of frozen products presents an entirely new set of requirements. In addition to the usual demands placed on the package, one must also take into account the effect of sub-freezing temperatures on the packaging materials as well as on the packaged product. One must furthermore consider the product in its frozen state and its demand on the packaging material. Lastly, one must pay attention to the mode of distribution and the ultimate utilization of the product by the consumer.

CASE STUDY 21-1

A packer of a pizzalike product had, due to large demand, switched his production to a horizontal form-fill-seal machine. The finished plastic pouch was then loaded into a printed carton. To facilitate this final packaging step, a partial vacuum was drawn during the sealing operation. Vacuum was not required to preserve the product. Freezing, after all, provided extended shelf life. However, the residual air in the package hindered the rapid insertion of the ballooning package into the box. By removing most of the air from the package, the loading process was immensely facilitated.

The introduction of this new packaging system was beset by all sorts of problems. This situation is often encountered with startup operations. However, after most difficulties were relegated to occasional annoyances, there remained one persistent problem—high leaker rate. The seals failed frequently and unpredictably.

Finally, after lengthy observation, the failures were traced back to a very common problem. In the product loading, as a last step, sauce was squirted on top of the item. The nozzle dripped and at times a drop landed in the seal area. When the seal was made, the contaminated area would not seal well and failed shortly thereafter. Once this was ascertained the corrective action was swiftly instituted. A dripless nozzle and a catch pan virtually eliminated the problem.

ADDITIONAL READING

Eller, D. 1986. Why fresh isn't always best—the truth about frozen vegetables. *Mademoiselle* **92**(11): 233.

Johnston, D. D. 1985. Winter salads from the summer garden. *Organic Garden* **32**(1): 104–106.

McWilliams, P. G. 1988. The great vegetable freeze. *Country Journal* **15**(8): 27–32.

Ross, V. 1976. Recent advances in packaging systems reflecting changing needs of frozen food industry. *Quick Frozen Foods* **38**(12): 21–24.

Showell, D. C. 1974. Problems in packaging frozen foods. *Food Processing Industry* (UK) **43**(511): 30–32.

Sison, E. C. et al. 1973. Acceptability of microwave reheated precooked chicken after packaging, freezing and storing. *Poultry Science* **52**(1): 70–73.

Spalding, B. J. 1987. A high tech answer to soggy frozen food. *Chemical Week* **140**(16): 30.

Winter, F. H. 1971. Preserving frozen-food quality. *Today's Packer* **1**(7): 12–14.

Q 22: Which Would Be the Most Suitable Package for Seafood?

A 22: The seafood industry has undergone some very startling changes and yet has in some respects remained the same.

SUPPLY AND DEMAND

For centuries, seafood has been a household staple, served up at least once each week. In some homes and in many restaurants as well it was and still is considered a poor meat substitute or a cherished delicacy, which some hated and others considered a gourmet treat. Millions of Catholics were forced to eat fish each Friday and some resented it. However, when this restriction was lifted, the consumption of seafood did not diminish. Quite the contrary, the per capita consumption in the US has gone from 11.8 lbs in 1970 to 14.4 lbs in 1985, a 22% increase in 15 years. Some claim that the retail value of seafood in 1989 was about $28.8 billion. In the last few decades while the demand for seafood has steadily increased, the supply seems to have decreased. If 14.4 lb/per capita is an accurate statistic—and there is no reason to question it—then the annual consumption for the USA should be 3.6 billion lb (14.4 × 250 million people). At an average retail price of $8/lb this would confirm the $28.8 billion of retail sales. However, the US "catch" for fish and shellfish for 1988 is quoted as 10.9 billion pounds. There seems thus little justification for ever increasing prices based on alleged shortages.

THE HEALTH ASPECT

Seafood is a ready source of protein and many other nutrients. Many fish oils are low in saturated fats and cholesterol. They are therefore a choice of the health conscious consumer. However, seafood has also received a massive dose of negative publicity. Early in the century there persisted the axiomatic rule. "Don't eat fish in months without 'R'." This meant that fish were tabu from May to August of each year. There was a good reason for this odd rule. Transportation and refrigeration was not up to today's standards. Thus seafood would spoil on the way to the retail establishment and could actually present a health hazard to the ultimate consumer. However, under present circumstances there is no reason to hold on to this outmoded belief.

Seafood has in recent times suffered from some very bad press. As the pollution problem worsened, we became aware of fish absorbing waterborne pollutants. Thus we had a mercury scare in the 1970s. Recently there was concern for bluefish contaminated with PCB or swordfish with mercury. Shellfish are even more susceptible to pollution since they are grown in beds just offshore. Unfortunately, there is little government control over seafood. Meats are subject to both FDA and USDA inspection. There are, however, no federal inspection authorities for fish. Local and/or state inspection of shellfish is not very effective. One precaution that every consumer can and should take, is the avoidance of raw fish. Well cooked seafood will have practically all microbial contamination destroyed. There are many examples of eaters of sushi and other raw fish or shellfish dishes suffering dire health consequences. Aside from the microbial problems, it has recently been demonstrated that worms and parasites can be transmitted as well.

A highly publicized case is that of "Kapchunka." This delicacy is a salt cured, air dried whitefish. This fish was sold uneviscerated (whole with its insides intact). The intestines harbored *Clostridium botulinum* and claimed at least one fatality, and several were taken very ill. A very fast reaction by FDA tracked down and recalled the product within just a few days and this held the damage to a minimum.

There are two types of diseases associated with seafood:

1. Scrombroid poisoning. This may cause serious discomfort but is not life threatening.
2. Ciguatera. This is primarily transmitted by reef dwelling fish in subtropical climates. Of commercial importance are barracuda and red snapper. The toxin is not destroyed by freezing, may attack the nervous and respiratory system, and could cause death.

FROZEN FISH

Food can be preserved for extended periods of time in a freezer. Any food market has rows of freezers containing a range of products. Seafoods have joined the broad spectrum of raw and prepared foods, readily available. It must be emphasized that freezing retards the reproduction of microbial organisms. However, whatever germs or toxins were present prior to freezing will survive during the freezer storage. What is worse, the microbes will resume their reproduction during the thaw cycle. It is essential to cook the frozen seafood well.

The reluctance of the consumer to purchase frozen meat has been noted (Q 28). This same resistance has been felt in the seafood department. The buyer is still after "fresh" fish. Frozen fillets or prepared dinners are finding some acceptance, though.

PACKAGING

Seafood is a very delicate product, subject to rapid spoilage. Thus any means of increasing the shelf life is desirable. Ice, refrigeration, salting, and smoking have all been employed to extend the storage time of seafood. Controlled atmosphere packaging should be able to make some significant contribution in this area. Gas or vacuum packaging has changed the shelf life of frankfurters from the unprotected less than 10 days to about 45 days. It is reasonable to expect that the same could be achieved for seafood.

Many benefits would spring forth from this packaging innovation. Currently, one depends on the "catch of the day" for supplies. With a 45 day shelf life, more variety would be available at all times. This in turn would eventually reduce retail prices. Since some fish farms have sprung up in the last few years, this mode of marketing would be most suitable for such operations.

The consumer, however, is still reluctant to accept prepackaged seafood. *Clostridium botulinum* is the dreaded invisible enemy and it can survive in vacuum or inert gas. The best defense against this horror is oxygen. Therefore, CA packaging has not been promoted extensively. More correctly, everyone has avoided even suggesting it. However, several developments may combine to give CA packaging of seafood a new chance. It has been found that the harmless lactic acid bacteria retard the growth of *Clostridium botulinum*. The former have actually been employed to extend the shelf life of shrimp. Since the lactic acid bacteria are facultative (they thrive in air or vacuum) they could be seeded in a CA package to retard the growth of the harmful microbes. Temperature control is another important factor. The useful retail life of the product can be extended manifold with proper cooling. Seabrook Farms has entered the field some years ago with liquid nitrogen freezing. However, more recently a CA package has made its appearance, as well. The fish is frozen and packed on ffs equipment in a CA package. At the retail level the product is permitted to thaw. The package is displayed refrigerated with an oxygen permeable membrane permitting oxygen to circulate through the package. Thus the best of all possible situations is achieved. The fish is quick frozen at the source and is therefore assured a long storage life. At the retailer the fish is displayed as (or like) fresh in an appealing package. The oxygen level in the package prevents *botulinum* problems.

Will the package sell? It is too early to give an answer.

ADDITIONAL READING

Adams, J. P. 1983. Processing of seafood in institutional-sized retort pouches. *Food Technology* **37**(4): 123–127.

Anon. 1990. Seafood entree enters shelf-stable market. *Packaging* **35**(4): 67.

Anon. 1988. Three packaging elements combine for seafood CAP. *Food Engineering* **60**(10): 53–54.

Anon. 1988. Unique fresh fish shipper wins AIMCAL award. *Packaging* **33**(3): 20.

Anon. 1987. Mackerel works miracles with blood pressure. *Prevention* **39**(7): 14.

Ballentine, C. 1986. Weighing the risks of the raw bars. *FDA Consumer* **20**(9): 20–23.

Higashi, G. I. 1985. Foodborne parasites transmitted to man from fish and other aquatic foods. *Food Technology* (3): 39–69.

Iverson, E. S. 1989. Sushi, sashimi, and sickness: raw fish and parasites. *Sea Frontiers* **35**(5/6): 176–183.

Mallozzi, A. C. 1990. How safe is our seafood? *Good Housekeeping* **210**(5): 253–254.

McGinnis, J. A. 1983. Raw food can be a hazard food. *Environmental Health* **45**(1): 182–185.

Modeland, V. 1989. Fishing for facts on fish safety. *FDA Consumer* **23**(2): 16–21.

Moon, N. J. et al. 1982. Evaluation of lactic acid bacteria for extending the shelf-life of shrimp. *Journal of Food Science* **47**(9): 897–900.

Segal, M. 1988. Fish "delicacy" causes botulism illness and death. *FDA Consumer* **22**(5): 33–34.

Q 23: How Should Fruits and Vegetables Be Packaged?

A 23: Produce is a very perishable commodity which is subject to a high waste factor. The techniques applied in post-harvest handling, storage, and market display are largely responsible for this high toll. While great advances have been made in improving the yield and speed of agricultural production, there has been relatively little done to extend the shelf life of fresh produce on a commercial scale.

The most interesting research on produce preservation has been conducted with controlled atmosphere storage. This technique has been highly successful in the laboratory and is now commercially applied to a very few produce varieties.

PACKAGING NEED

A need for reducing the rate of spoilage encountered in the marketing of fruits, vegetables, and cut flowers is self evident. The losses to the farmer, the wholesaler, the retailer, and the consumer are staggering. The spoilage factor discourages the production, distribution, and consumption of many of these vegetable products. A protective package may actually contribute to a lowering of cost (not profits) and a more plentiful supply of fruits, vegetables, and flowers. The market size (Table 23–1) is certainly adequate to justify packaging considerations.

PACKAGING COST

The cost of an improved produce package is well within the range of established packaging cost for other food items. The ratio of product to package

Table 23-1. Produce Market Size, U.S. per Capita Consumption, lb.

	1975	1980	1985
Fresh fruit (retail)	81.5	86.4	86.3
Select fresh vegetables	66.6	72.8	78.8
White potatoes, fresh and frozen	68.0	69.9	66.7
Bananas	17.6	20.8	23.4
Apples	18.2	18.3	16.6
Grapes	2.9	3.3	6.6
Oranges	15.4	15.4	12.0
Lettuce	21.9	24.9	23.2
Tomatoes	10.2	11.4	13.7
Onions	12.6	12.9	15.6

cost for processed meat is approximately 36. For grapes, strawberries and mushrooms the ratio would be about 33–46. For cut flowers the ratio might run from 25 to 40. The increased shelf life and reduced spoilage will more than pay for the package cost. Not all produce items, however, will be able to carry the increased package cost.

SPOILAGE

The one factor that accounts for relatively high prices of many produce items—the one factor that limits production and precludes broader national and international distribution—that one factor is spoilage. Many fruits and vegetables survive for just a few days even under the best conditions (See Table 23-2). The high price of strawberries (about 79¢ for 8 to 10 ozs.) and the limited seasonal availability are no doubt related to the maximum 5-day shelf life. Supermarkets carry strawberries as a service to their customers rather than a profitable sales item.

Similarly, cut flowers have a very high rate of spoilage. Many supermarkets carry cut flowers for a variety of reasons. All agree that a large volume could be sold once the spoilage factor could be brought under control. The rates of spoilage of some select popular flowers are shown in Table 23-3. The reasons for such rapid spoilage are manifold. Most vegetation would simply shrivel up from loss of moisture. To overcome this problem, cut flowers are placed in water (preferably warm water), lettuce is overwrapped with cellophane, and a variety of other fruits and vegetables are placed in polyethylene bags. Moisture retention, while a prime concern, is by no means the panacea for spoilage.

Respiration

Breathing continues even after plants are separated from their roots or stem. Thus flowers or produce will continue to inhale oxygen and exhale

Table 23-2. Storage Life of Produce.

| Produce | Recommended Conditions | | |
	Temperature (°F)	Relative Humidity (%)	Duration
Avocados	40–55	85–90	2 weeks
Papayas	45	85–90	1 week
Figs	31–32	85–90	1 week
Strawberries	32	90–95	5 days
Raspberries	31–32	90–95	2 days
Broccoli	32	90	10 days
Corn (sweet)	32	90–95	4 days
Cantaloupe	32–35	85–90	5 days
Mushrooms	32	90	3 days
Okra	45–50	90–95	7 days
Tomatoes	45–50	85–90	4 days
Watercress	32–35	90–95	3 days
Carrots	32	90–95	4 months
Pears	29–31	90–95	7 months

Table 23-3. Storage Life of Cut Flowers.

| Flower | Recommended Storage | |
	Temperature (°F)	Duration (days)
Daffodils	32–33	10–21
Daisies	40	3
Forget-me-not	40	1–3
Gladiolus	35–50	6–8
Lilac	40	4–6
Roses	32	7
Violets	33–40	3
Tulips	31–32	4–8 weeks

carbon dioxide from harvest to the time of consumption. The plant is subject to spoilage by suffocation or anaerobic oxidation.

Rot

The vegetative products are subject to infections. Bacteria, mold, and a variety of other microscopic organisms' are ever present and ready to attack the vegetative tissues. Often these microbial attacks spread rapidly through the entire harvested crop. Certainly contact with infected surfaces promotes the transmission of the disease to healthy plants. In the early stages of infection there may be minor blemishes. This is often followed by flavor alteration, decay, foul odor, etc.

PRESERVATION

There are several proven methods of extending the useful shelf life of produce. A combination of these is probably required to reach the optimum in preservation. Unfortunately, each plant has its own unique survival requirements and it is thus impossible to design the universally acceptable "best" conditions.

Temperature

Low temperature reduces the respiration rate and also retards the growth of microorganisms. Thus by placing cut flowers, vegetables, or fruits under refrigeration one reduces the chance for spread of infection while holding down the rate of aging. Each plant has its own optimum storage temperature. Some vegetation is susceptible to chilling injuries. Thus bananas should not be stored below 53° (do not place the bananas in a refrigerator), cucumbers below 45°, grapefruits below 50°, tomatoes below 55°, etc. Storage at lower temperatures may lead to discoloration, off flavor, and even rot. In practical terms this variant in temperature tolerance makes it very difficult to design a suitable storage or retail display case to handle all the various vegetative products.

Moisture

The relative humidity in the intercellular spaces of plant tissues is 97–100%, whereas the outside atmospheric humidity is usually much lower than that. There is a constant gradient of vapor pressure that causes water from the fruit or vegetable to diffuse outside the product, thus resulting in shriveling. To eliminate this problem, a fairly high relative humidity is desirable in storage rooms. Caution should be exercised to control moisture levels, as excesses of water vapor in the surrounding atmosphere can lead to condensation on the produce surface and stimulation of mold growth. Low temperatures help to retard mold growth but a good practice is to permit a slight drying of the product so that water vapor is allowed to leave the immediate proximity of the produce.

Handling and Disease

Many fruits and vegetables are tender and subject to physical change if shocked. Precautions should be taken in handling the produce during harvesting, packing, and shipping so that the products do not arrive at the market badly bruised. Close attention should be given to the inspection for diseased produce during pre-shipment packing. The diseased or mold-infected product can easily spread its infection to other items in the shipping case and thus ruin much of the contents. Subdividing bulk shipments into hermetically sealed portion packs may isolate and prevent the spread of an infection from one contaminated area throughout the entire shipment.

Controlled Atmosphere (CA)

Another means of prolonging market life of fresh produce is to slow down respiration by modifying the storage atmosphere. There is general agreement that decreased levels of oxygen and increased levels of carbon dioxide retard softening, color development, and aging, and sometimes inhibit decay development in stored produce. Generally speaking, a reduction of the concentration of O_2 to 5% is usually beneficial in retarding ripening.

It should be stressed, however, that each fruit or vegetable has an optimum level of concentration for maximum storage life. An atmosphere that benefits one kind of produce, or even one variety, may not benefit another. In some instances modified atmospheres cause injury and off flavors. If oxygen levels drop too low, anaerobic respiration and fermentation can occur. Carbon dioxide generated during respiration, can accumulate to injurious levels. The atmosphere in storage areas must at times be scrubbed to keep CO_2 content at predetermined volume. The scrubbers absorb CO_2 and can be chosen from such substances as calcium hydroxide, sodium hydroxide, ethanolamine, or even water.

Some fruits and vegetables generate ethylene gas during the ripening process. This fact is taken advantage of in picking green tomatoes and adding ethylene gas during the storage period. The green tomatoes will turn pink at the desired marketing date. However, ethylene gas may continue to ripen the produce beyond the desired stage of development and thus shorten its useful shelf life. Control of ethylene evolution or its rapid dissipation can contribute to improved product survival.

Controlled atmospheric research has been tried on a good number of produce types. Most testing has involved pre-market storage, as opposed to retail market shelf storage. After harvesting, the fruits, vegetables or flowers are stored in automatically controlled environments of optimum temperature, carbon dioxide, oxygen, and nitrogen levels. Each produce type has individual gas and temperature requirements and so must be stored in a different environment. When the produce is ready for retail sale, it is removed from CA storage and placed on display at room temperature, usually with the result of extended display life. The atmospheric requirements of various fruits, vegetables, and flowers will be reviewed here to get an idea of the wide variety of atmospheric conditions that are needed to preserve different types of produce.

FRUITS
Apples

One of the few successful commercial applications of CA storage has been experienced with apples. The greater the depression of respiration of apples in CA, the greater the beneficial effects on shelf life after storage. The storage requirements for apples differ according to variety, but in general they are stored at 30–38°F in 2–3% O_2 and 1–5% CO_2.

Pears

Pears, particularly Bartletts, are best preserved at about 32°F in 0.5–1% O_2 and 5% CO_2 for early harvested fruit. For later harvested fruit 32°F in 0.5–1% O_2 and no CO_2 is preferred.

Bananas

The best atmosphere for banana storage is 2% O_2 and 2–7% CO_2 at 59–60°F and high relative humidity. The high humidity is important, as green bananas will fail to ripen below 85–90% RH. Unfortunately, the use of CA for bananas is both expensive and cumbersome. Investigations have now turned to vacuum packaging of bananas in polyethylene with favorable results.

Avocados, Mangos, and Limes

Avocados were stored for about 30 days at 40°F in 7% CO_2 and 1% O_2. Mangos can be stored at 55°F in 5% O_2 for 20 days. An O_2 level of 1% was found to cause off flavors and skin discoloration in mangos. Limes can be stored for 60 days in 7% CO_2 and 5% O_2 at 50°F.

VEGETABLES
Tomatoes

Tests indicate that mature green tomatoes can be held longest in an atmosphere of 55°F, 3% O_2, and 97% N_2. They remained firm and green for up to 6 weeks and then ripened normally with acceptable flavor when removed to air at 65°F. There is some indication that tomatoes attain better color when ripened at low relative humidity.

Cabbage and Brussel Sprouts

Temperature is very important in preserving stored cabbage and brussel sprouts. Both vegetables should be stored at 32–34°F and over 96% relative humidity. The atmosphere should be 5% CO_2 and 1% O_2.

Cauliflower

Cauliflower stores best in air at 32°F for up to 8 weeks. CO_2 concentrations of 5% proved detrimental to appearance of the product.

Asparagus

Asparagus can be kept 7 days at 37–43°F in 5–10% CO_2. The CA gives better market quality than air and reduces the occurrence of bacterial soft rot.

Mushrooms

Mushrooms have a very short shelf life. When stored at about 3°C (38°F) they will only last for 3 days. Gas treatment helps to extend storage life considerably. An atmosphere of 100% nitrogen at 3°C increased the shelf life to 6 days. When treated with nitrogen then carbon monoxide, the mushrooms remained satisfactory in quality for 15 days.

CUT FLOWERS
Daffodils

Temperature is a critical factor for storage of any cut flowers. Tests on cut daffodils stored in air without water revealed a storage life of 1–2 days at 80°F, 2–3 days at 70°F, 3–5 days at 60°F, 6–8 days at 50°F, 8–15 days at 40°F and 21–28 days at 32°F. Daffodils stored in 100% N_2 and 32°F had the best display life of all atmospheres. After 3 weeks at 32°F in 100% N_2, daffodils were removed to a 73°F atmosphere and had a display life of 103–123 hours in contrast with 85–98 hours when stored in air and 90 hours when stored at 40°F in 100% N_2.

Gladiolus

Storage in 1% and 5% CO_2 for 6–8 days at 32° or 40°F, respectively, was slightly beneficial in aiding floret openings. Those flowers stored in low O_2 with 5% CO_2 were appreciably better than 0% CO_2. More upper florets partially opened after removal from CA, whereas those stored in air dried and wilted.

Carnations

Storage at 32°F was satisfactory for 2 weeks or less but caused petal injury to red and pink carnations during longer periods. It was found that 36°F was best for longer storage. Carnations were best stored dry in 0.5–1% O_2 for 4–5 weeks at 36°F.

Roses

Dry storage of roses for 2–3 weeks was better at 32°F than at 36°F. In general, most CA treatments were not beneficial in extending storage life at 32°F. An atmosphere of 0.5% O_2 and 5% CO_2 was best of those tested.

CONCLUSION

1. Each variety of produce has its own optimum storage environment and consequently must be dealt with individually for packaging consideration.
2. Storage temperature is of critical importance in preserving produce. Temperatures should be held to the lowest point at which freeze damage does not occur.
3. Produce packaging films should have sufficient water vapor barrier to retain product moisture without accumulating excessive moisture to support mold growth.
4. Produce packaging films should have CO_2 permeability of about 3000–12000 cc/24 hours × 100 sq. in. × atm, so that CO_2 does not accumulate to injurious levels.
5. Produce packaging films should have a relatively low O_2 permeability of 700–2500 cc/24 hours × 100 sq. in. × atm so that produce respiration is sufficiently retarded without creating anaerobic damage.
6. Produce packaging should provide structural protection against shock damage and separation to prevent spread of mold growth and disease.

ADDITIONAL READING

Anon. 1988. Extended fruit and vegetable shelf life via permeable films. *Food Engineering* **60**(2): 39–41.

Anon. 1985. What makes petunia flowers wilt? *New Science* **106**(4): 22.

Anon. 1985. Produce packaging plant uses anaerobic technology and hybrid trees to solve effluent odor problems. *Water Pollution Control* **123**(3/4): 72.

Anon. 1977. Vacuum-packed shredded lettuce stays fresh for 12 full days. *Food Engineering* **49**(4): 81–83.

Anon. 1976. Effects of vacuum packaging on quality and storage life of broccoli. *Lebensmittel-Wissenschaft und Technologie* **9**(8): 251–254.

Anon. 1975. Prepackaging of fruits and vegetables in plastic films. *Neue Verpackung* **8**(2): 127–134.

Bhowmik, S. R. and Sebris, C. M. 1988. Quality and shelf life of individually shrink wrapped peaches. *Journal of Food Science* **53**(2): 519–522.

Harvey, J. M. et al. 1988. Sulfur dioxide fumigation of table grapes. *American Journal of Enology and Viticulture* **9**(2): 132–136.

Kader, A. A. 1986. Potential applications of ionizing radiation in postharvest handling of fresh fruits and vegetables. *Food Technology* **40**(6): 117–121.

Larson, M. 1990. MAP/CAP satisfies craving for freshness. *Packaging* **35**(9): 68–70.

Smith, S. M. et al. 1988. The effect of harvest date on the response of Discovery apples to MA retail packaging. *International Journal of Food Science and Technology* (UK) **23**(1): 81–90.

Watada, A. E. 1986. Effect of ethylene on the quality of fruits and vegetables. *Food Technology* **40**(5): 82–85.

Yeaple, J. A. 1989. Putting fruit to sleep. *Popular Science* **234**(5): 167.

Zagory, X. and Kader, A. A. 1988. Modified atmosphere packaging of fresh produce. *Food Technology* **42**(9): 70–74.

Q 24: What Materials Are Best Suited for Snack Food Packaging?

A 24: To better understand the needs of snack food packaging one must review the type of snack foods available and the industry that produces this product.

SNACK FOOD PRODUCTS

The term "snack food" applies to a wide range of items usually consumed between regular meals. Usually these are food items not normally required for sustenance. A wide variety of products are broadly classified as snacks. More than 60% of all snack foods sold are either candies, crackers, cookies, or potato chips. Figure 24-1 depicts nine broad classifications and the sales volume (in dollars) ascribed to each. Under the "other" are included categories each accounting for less than 3% of the total sales volume, such as dried fruits, extruded snacks, fabricated chips, granola snacks, hot snacks, meat snacks, pretzels, and toasted pastries. However, this segment is growing faster than other categories, with the exception of popcorn. Popcorn increased almost eightfold, while "all others" consumption over the past ten years gained almost fourfold in sales.

It is important to remember that under each classification there are a great many varieties and each of these in turn may require a special packaging solution. Potato chips, for example, are packaged at a moisture content of 1.0–1.5%. When these chips reach a 4.0–4.5% moisture content they are considered stale. Meat snacks on the other hand or pretzels, or frozen snacks have an entirely different preservation problem.

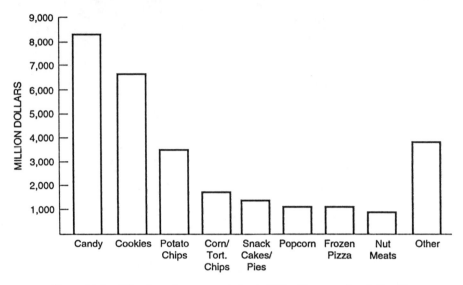

Figure 24-1. US sales volume—snack foods 1989. (*Courtesy Snack Foods*)

MARKET GROWTH

Snack consumption has almost doubled in the past ten years (Table 24-1). The rate of growth of this segment of the food market has been disproportionately high. Some ascribe the rapid rise of snack foods to a changing lifestyle, increased outdoor activity, more TV viewing, the singles scene, etc. No doubt the proliferation of types of snack foods contributes to the larger consumption, as well. The total sales volume for snack foods exceeds 29 billion dollars annually. A very conservative estimate for potential packaging material requirements would have to place same in excess of 400 million dollars. A distribution of packaging material uses is shown in Figure 24-2.

DISTRIBUTION AND MERCHANDISING

A large segment of the snack food industry distributes through driver-salesmen. Store personnel is thus relieved of all responsibility related to stock maintenance, reorder, inventories, etc. The driver has a certain shelf space assigned to him and he maintains merchandise on that allotted display. The cost of this personalized service is no doubt high. However, many snack food manufacturers will argue that this mode of distribution is not just the preferred but actually the one and only means of getting a fragile product into the supermarket and maintaining fully stacked shelves. This driver-

Table 24–1.   U.S. Sales Volume (million $), Nine Major Snack Food Categories, 1980–1989.

Product	1980	1981	1982	1983	1984	1985	1986	1987	1988	1989
Candy	4684	4939	5300	5625	6150	6683	7000	7430	7782	8301
Cookies/crackers	3663	3960	4197	4449	4938	5234	5627	5820	6214	6780
Potato chips	1940	2130	2289	2430	2514	2651	2850	2890	3048	3275
Corn/tortilla chips	840	1029	1132	1214	1246	1283	1324	1421	1607	1847
Snack cakes/pies	885	982	1000	1000	1110	1220	1240	1364	1429	1466
Popcorn	135	153	160	163	176	365	458	801	993	1100
Frozen pizza	730	774	759	774	798	864	918	974	1000	1087
Nut meats	1089	980	1088	1099	1121	1160	1160	1098	1050	974
All other	1077	1495	1694	1935	2285	3049	3509	3807	3905	4236
Total	15,043	16,442	17,619	18,689	20,338	22,509	24,086	25,605	27,028	29,066

Source: Snack Food, June 1990.

FLEXIBLE PACKAGING CONSUMPTION
SNACK INDUSTRY - 1989

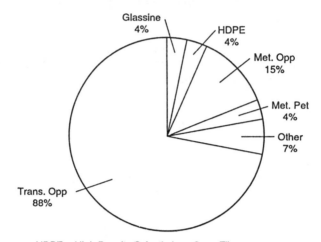

HDPE = High Density Polyethylene Coex Film
Transp. Opp = Transparent Oriented Polypropylene Film
Met. Pet = Metallized Sealable Polyester
Met. Opp = Metallized Opp
Glassine = 25 lb. basic wt. with 5 lb. PVDC coating

Figure 24-2.   Flexible packaging consumption. (*Courtesy Snack Foods*)

salesman approach is credited with maximum sales and minimum returns. It is further claimed that because of this distribution approach packaging costs can be minimized. Even relatively inexpensive packaging materials will survive the rapid turnover achievable with driver-salesman. Cookies that are distributed through normal warehousing channels must go to expensive foil overwraps and still encounter a high percentage of breakage. The economics of the driver-salesman distribution is by no means as clear cut as some make it out to be. In recent years a number of companies have entered the market with new snack foods distributed through the warehousing route. They have been very successful with this marketing and distribution approach. Unfortunately, no figures are available from either side on return rates and one can thus not develop a total distribution cost analysis. Certainly the supermarket looks with favor upon snack foods. They are fast moving, high profit items. Given a choice, the supermarket prefers the driver-salesman distribution since it relieves the market manager of any and all responsibilities for these items. However, it is to be assumed that were snack manufacturers to change to warehousing distribution, the supermarket would continue to display and sell these items.

GEOGRAPHIC PREFERENCES

There seem to be differences in type and kind of product consumed depending on regional preferences. Even the same products, such as potato chips for example, may be marketed in a slightly different form depending whether they are sold in New England, Texas, Illinois, or Oregon. Along with the differences in product shape, appearance, and taste there is also a distinct difference in the packaging approach. This may well explain why so many regional snack food packagers flourish.

THE PACKAGE

There obviously is no single package to encompass so many different products, nor even a consensus for packaging a single product uniformly throughout the country. One must therefore look at the various segments of the market and decide what some of the prevalent packaging trends are.

Candy

Most hard candy is packaged in polyethylene or polypropylene bags to provide minimal moisture protection. The product is relatively inexpensive and requires little shelf life. There are, however, more expensive candies that can justify a more sophisticated package. Many of these have in recent times switched from a foil to a metalized or coextruded composite. The new material is not only less expensive but can be sealed on much higher-speed packaging machines.

Crackers and Cookies

A great many of these are still packaged in boxes. In some instances they are poly-coated chip board without an inner liner. However, most do have a plastic bag in the box. The bag may be polyethylene or PVDC coated glassine. Consumers like this approach because it permits the reclosing of the package after extracting a few cookies. Another approach has several smaller packages of crackers or cookies within the box. Thus only a small portion of the content is opened and if not rapidly consumed, just a segment of the contents goes stale. Another favorite packaging mode for cookies is a square-bottom pouch. It, too, permits the consumer to reclose the bag in order to preserve the remainder. Less costly cookies are normally packed on a styrene tray with a cellophane or polypropylene overwrap. The package does not offer a high degree of protection. However, due to its local distribution, the need for product protection is minimized. The package does provide good product visibility and with its printed design affords product recognition. It is surprising that in this huge market, estimated to exceed 6.5 billion dollars worth of sales, no one has considered a modern packaging approach. The box, with or without the liner or the overwrapped polystyrene tray, is a most primitive way of packaging. None of these packages face up to the possibility of rancidity, moisture loss, and breakage, all of which must be responsible for an appreciable return rate. But then, this is how the cookie crumbles.

Potato Chips

There seems even less uniformity in packaging material usage for this type of snack food. This product contains a high fat content and is thus subject to flavor deterioration brought about by rancidity. Oxygen reacts with unsaturated fats in the presence of UV light to produce peroxides which are the precursors of rancidity. The ideal potato chip package would thus exclude oxygen as well as light in order to achieve prolonged shelf life. At this writing there is no chipper using a low oxygen atmosphere to protect his product. The Waldman and Mira Pak machines which are predominantly used in this packaging operation are not equipped to provide a gas package. Obviously, a vacuum package cannot be produced since the product is too fragile and will crumble under these conditions. Some chippers have gone to an opaque package to reduce the UV exposure of the product and thus minimize off flavor. However, other chippers feel that the visibility of the product increases its sales appeal and therefore package in transparent materials in order to display their product. More important than the oxygen problem is that of moisture absorption. A packaging material is considered suitable if it permits less than 3% moisture absorption during a specified shelf life period. Thus a local chipper whose product will be sold within a two to three week period may be satisfied with a single-ply cellophane or polypropylene. This same packaging material will not suffice for a national distributor who must obtain twelve weeks' shelf life. Traditionally, the small chip package has been provided with an inexpensive packaging material such as PVDC coated glassine or a single-ply PVDC-coated film. The larger package on the other hand has usually had more sophisticated packaging materials assigned to it. The chipper, however, has failed to realize that he faces a paradoxical requirement, namely, that the smaller package requires the more costly packaging material. If one considers moisture protection only (and there are many other factors that go into the selection of a suitable packaging material) then the smaller package requires better moisture protection. The ratio of square inches of material per ounce of product drops drastically the larger the package becomes. For example, in the small 3/4 ounce package one utilizes 64 square inches per ounce of product while in the large family size package of 7-1/2 ounces one requires only 16 square inches per ounce of product. It is thus apparent that one can utilize a packaging material that has a fourfold increase in moisture vapor transmission in the larger package over that in the small package.

Cellophane was a favorite packaging material with potato chippers for many years. It provides sparkle and a certain degree of "body" to the package. Thus the package maintains a good rigidity while sitting on the shelf. Its crinkle, too, is a sound that conveys freshness to the consumer. It has many other advantages too, such as printability, sealability, and certainly not least is its ability to tear readily and thus provide an easy open mechanism for this package. Cellophane, however, has disadvantages as well. The most troublesome in the past has been its poor performance at low temperatures and its limited shelf life. Cellophane will dry out under low humidity conditions and embrittle and cause package failure. The most damaging blow to cellophane, however, has come in recent years with a rapid increase in price. This has brought other plastics into play as potential replacements.

Polypropylene has replaced cellophane to a large degree. These two films were used for many years in combination with one another. Thus a ply of cellophane was laminated to a ply of polypropylene. For a while there was a debate as to whether the polypropylene should be on the outside and thus protect the cellophane from changing weather conditions or whether the polypropylene should be on the inside offering better abrasion and puncture resistance to the product. However, in recent times there have been changes eliminating cellophane to a large degree. Newer packaging materials consist of "all plastic" composites. This involves two plies of polypropylene laminated to one another or an outer ply of polypropylene with an inner ply of high-density polyethylene. Metallized polypropylene and polyester have gained a degree of acceptance, as well.

Cellophane is attempting a comeback based on environmental considerations. It is claimed (Table 24-2) (Banerjee et al.) that cellulosic films, even if PVDC coated, are biodegradable. Cellophane and plastic films were buried in soil and examined for up to one year at regular intervals. The cellophane totally disintegrated within 10–84 days. The longer interval was noted with PVDC and vinyl coated film. By way of comparison, it took maple leaves 84–112 days to fully disintegrate. Thus the coated cellophane is no worse than nature's own leaves. Plastic films such as PVC, OPP, or PET did not show any signs of degradation at the end of one year.

Sealing of these packages has always been a serious problem. On the machines mentioned earlier and others as well, the packaging material is formed into a tube. The back seal, the one along the longer edge of the package, is formed by either sealing the back to the front (lap seal) or the

Table 24-2. Soil Disintegration Time.

Film Type	Gauge	Disintegration Time (days)
Uncoated cellophane	90	10–14
PVDC coated cellophane	80	28–56
Vinyl coated cellophane	100	56–84
Polyester	142	*
Polyvinyl chloride	150	*
OPP	100	*
Maple leaf	—	84–112

*No detectable weight loss after one year burial.

same surfaces to one another (fin seal). If the composite has PVDC on both surfaces (outer and inner) then either of these seals can be achieved by sealing PVDC to PVDC. The seals are rather weak but adequate for the relatively low-density products packed. To reduce packaging cost there has been a tendency to eliminate at least one of the PVDC coatings. This still permits formation of a fin seal but excludes lap seals. In order to enable the chipper to continue formation of lap seals, thermal stripe was introduced. This in essence is a thin stripe of adhesive applied at the edge of the outer surface. When the tube is formed the inner surface seals onto the preapplied thermal stripe. The thermal stripe is appreciably less expensive than the overall PVDC coating and yet permits the chipper to retain his old packaging operation. Others have eliminated the PVDC coating entirely and have gone to other heat seal media such as polyethylene, for example, and switched to fin seals to reduce their package cost even further.

Corn Chips

All the consideration of packaging discussed under potato chips are equally applicable to this product. However, it must be noted that corn chips are denser and have sharper edges. Thus there is a greater abrasion and puncture problem and packaging materials must be upgraded somewhat to allow for a potential package failure problem.

Nut Meats

A variety of nut products are sold in flexible packaging materials. A much larger portion is sold in glass or cans. The primary problem encountered is related to oils and fats commonly associated with nut meats. While glass and cans are vacuum packed, there is no well known brand in the flexible pack that is either vacuum or gas packed. Some small packages, especially those handed out by the airlines, are in foil composites or more recently in metallized polyester composites. The retail packages, however, were primarily in cellophane composites. Again, the many advantages of cellophane were recited under potato chip packaging. The oil problem, aside from rancidity, is one of potential transfer onto the packaging material, which would diminish its transparency. Rubber hydrochloride film (Pliofilm) seems to overcome this problem and has been used extensively in conjunction with cellophane as a suitable flexible nut package. Polyethylene and Surlyn have been finding their way into this packaging material more frequently as of late and polypropylene has essentially replaced cello in many applications. Metallized polypropylene or polyester make a very attractive package and provides a UV barrier at the same time. The price of such a pack is reasonable and it is finding ever increasing market acceptance.

CASE STUDY 24-1

An example of difficulty caused by deviation from good practice occurred at a converting facility after a recent overhaul and upgrade of equipment. A snack lamination specified priming of reverse printed cellophane followed by extrusion lamination to OPP with 7 lb/ream of LDPE extrudate. Upon starting up, bond in clear areas was found to be poor and in inked areas was virtually nil. Batch changes were made of the various components, yet the problem persisted. Ultimately, this job was removed and the next assignment begun. This work was similar to the earlier job but involved 15 lb/ream of extrudate. Quality of bond was perfectly good and normal. Upon seeing this, the operator was instructed to reduce the extrudate quantity stepwise in an attempt to arrive at the 7 lb/ream condition. This was attained with continual good bond. Upon further reduction of extrudate, bond at 6 lb began to fall off and became totally bad, as described above, at 4–5 lb/ream of extrudate. The engineer then checked weight of extrudate on the rejected production and found only 4–5 lb/ream despite having programmed 7 lb/ream. The difficulty was traced to a new beta ray gauge installed to measure and adjust extrudate thickness. The gauge appeared to be reading ink as if it were extrudate and kept reducing extruder output. At a point just below 7 lb/ream, the molten mass was allowing its heat to dissipate too quickly and not wetting out effectively on the base web.

CASE STUDY 24-2

A converter was setting up to assemble pouch stock composed of high barrier cellophane/PX/aluminum foil/coextruded heat seal. A single-head coextrusion line with both pre-prime and post-prime stations had to be employed for this job, which had to be accomplished in two passes.

The first pass combined the cellophane/prime/PX/foil/primer. The post-prime step is intended to prevent chemical attack of the foil surface by any components of the cello. After rewinding, the second pass applies the coextruded heat seal layer onto the primed foil surface.

As in any converting operation, turnaround time between passes might be scheduled for some maximum time limit, but often is delayed by processing or equipment problems. After just twelve hours in the rewind, less-than-optimum bonds in the finished structure were noted off-machine and subsequent package tests disclosed delamination. The problem was traced back to the jaw release coating on the cellophane, which offset face to back in the rewind, coating the primed foil with an antiadhesive layer.

The only assured way of dealing with this problem was to start this process with cello which was free of jaw release coating. Then, at the end of the second pass, a wash coat of surfactant jaw release was applied to the cello surface at the post-prime station. (For priming information consult Mica Corp; Stratford, CT).

ADDITIONAL READING

Avera, R. 1974. The renaissance of fin seals. *Snack Food* **64**(5): 42–43.

Banerjee, S. K. et al. 1970. How quickly do packaging films disappear down in dirt? *Package Engineering.* **15**(11): 211.

Broekel, R. 1989. The great American candy bar. *Antiques and Collecting Hobbies* **94**(3): 27–29.

Lindgren, B. J. 1988. Award judges ask "can flexo printing get any better?" *Flexo* **13**(5): 10–50.

Long, H. 1974. Proper packaging maximizes profits. *Snack Food* **63**(1): 58–60.

Prince, Ph. 1990. Barrier protection key to optimal snack packaging. *Paper, Film and Foil Converter.* **64**(9): 72–74.

Taylor, C. C. 1971. Flexible packaging for snacks. *Snack Food* **60**(1,2,3,4).

Q 25: Which Is the Best Bacon Package?

A 25: Bacon is a pork product with a very high fat content. It is sensitive to bacterial decay and should be protected by refrigeration as well as controlled atmosphere packaging. Nevertheless, a goodly portion of bacon is still packaged in tux cartons or simply over-wrapped. (Fig. 25–1)

Figure 25-1. Bacon in tux carton. (*Courtesy Armour & Co.*)

Meat ranks fairly high in the diet of the US population. Yet, this country is by no means the number-one meat consumer in the world. More beef and veal is eaten per capita in Argentina and Uruguay. In Saudi Arabia and Israel, the average person eats more chicken than the average American. Also, with respect to pork consumption, the US ranks no higher than twenty-second in the world. Fresh pork is much more perishable than beef or veal, and this must be considered when designing packaging for such products (see Q 28).

Both ham and bacon are cured pork meat. Ham originates from the hog's hind leg. The veins are injected with curing (pickle) solution, and then soaked in the same solution. The cured ham is thereafter "smoked" for about 24 hours. The ham can be "ready-to-eat" if the smokehouse temperature was high enough to cook the meat thoroughly.

Bacon comes from the sides (spareribs) of the hog. This product has a very high percentage of fat. This is one of the reason why pork in general, and bacon particularly, have gradually diminished in popularity. The American public has become very diet conscious, and "fat" has become synonymous with "unhealthy." Bacon, like ham, is first cured and thereafter smoked. The curing solution contains nitrates and nitrites as preservatives. The presence of these salts, too, gives some concern, due to the possible health hazards associated with these chemicals.

The vacuum package has many advantages over the older, cheaper modes of packaging. Bacon can survive for up to 60 days in a good vacuum package under proper refrigeration. The vacuum induces the packaging material to cling tightly to the product. This contact provides a very favorable display for the bacon. Because the vacuum package prevents motion of the product within the package, an orderly appearance of shingled bacon can be maintained.

Normally, bacon slices are arranged in a manner reminiscent of the shingles on a roof. Each successive slice covers most, but not all, of the previous slice. Thus a stack is generated which has a small portion of each slice exposed. The visible portion is by sheer coincidence the meaty edge of the bacon slice. The package thus exposes a high proportion of meat and hides the less attractive major portion of fat. The shingle has one other merchandising advantage. It creates the illusion of plenty. One of the leading bacon packers has merchandised an 8 oz. bacon shingle of extra thin sliced product arranged on a very large board. At first glance, one would have taken this attractive package to be a good buy, holding more bacon than is available in the standard package.

The shingle has to be transported from the slicer to a weighing station and from there to the packaging machine. This is usually accomplished on a carrying board. The carrier could be bought impregnated with wax or coated with polyethylene. In recent times legislation has mandated visibility of at least a large portion of one slice. Most packers prefer not to expose the top slice and have chosen to provide visibility on the back side. Thus plastic—primarily polypropylene—boards have made their debut on the scene. The traditional bacon board has been modified to include a "window" which meets the mandatory exposure of a portion of a slice (Figure 25-2).

In addition to its function as a carrier, the board provides body and stiffness to an otherwise limp product. Most packers utilize the board for transmittal of printed messages as well.

The shingle moving on a conveyor toward a packaging machine has a flat bottom and a curved top. To place the shingle into a form-fill-seal machine represents a unique problem. Most flatbed machines will form the bottom web. Thus it would be necessary to flip the shingle upside down in order to fit the formed top of the shingle to the formed bottom of the package. Attempts to introduce this system of packaging have not been very successful. The most accepted packaging equipment utilizes a wheel. The top web is formed on this wheel's dies and the shingle is introduced in its normal configuration into this preformed top web.

The most widely accepted of these machines was the FLEX-VAC 6–18, produced by Standard Packaging Corporation. A later modification of this

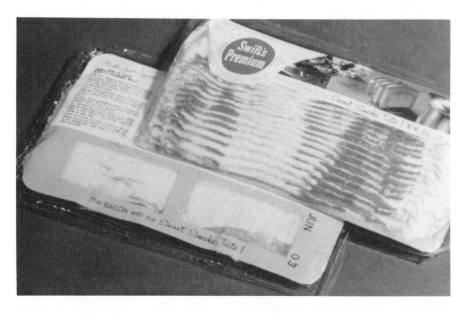

Figure 25-2. Traditional bacon package. (*Courtesy Swift & Co.*)

basic idea is also available as the Mahaffy & Harder 301 model. In recent times high-speed versions of these machines have been introduced to accommodate a higher production rate. The early models ran at approximately 45 packages/minute while the more recent ones can achieve approximately 100 packages/minute. The need for such high speed equipment is at this time debatable. Their utility will be established if and when high-speed slicers can be installed to keep up with the rate of the packaging machine.

A wide range of packaging materials are in common use at this time. Basically the packaging material consists of at least two and possibly three components. The outermost plastic material provides strength, abrasion, and impact resistance, as well as heat resistance to survive contact with hot sealing bars. The innermost component of the package is a sealant such as polyethylene, ethylene-vinylacetate copolymer, or ionomer, all of which provide for a hermetic seal under a set of time, temperature, and pressure combinations. Between these layers there is an optional third layer of PVDC which provides improved gas and moisture vapor barrier properties.

Packaging material selection must take into account the materials mentioned above as well as many other factors, including distribution range, size, quality of the product, the machinery on which packages are made, and many others. Selection of the most suitable material is best accomplished by materials supplier working in concert with the packer.

FORMED WEB

The material which is formed into a pocket to accommodate the shingled stack is normally a Nylon composite. Nylon is selected because of its superior thermoforming capabilities as well as its general strength characteristics. An economy pack will employ 3/4 mil (0.00075″) Nylon laminated to 2 mils of an olefinic homo- or copolymer. This material seems adequate and is used in large volume by some of the major national bacon packers. Quality houses, however, prefer to use a standard material of 1 mil Nylon and 2 mils of olefinic homo- or copolymer. This extra quantity of Nylon provides a margin of error which assures better package survival especially under rough handling conditions.

Thermoformable polyester can be substituted for Nylon especially where low-profile shingles are concerned. One bacon packer utilizes the formable polyester to print the formed web. The graphics on a formed web provide better display than that normally available on a board.

UNFORMED WEB

This side of the bacon package is normally composed of 50 ga.(12μ) polyester laminated to 2 mils (50μ) of an olefinic homo- or copolymer. It is important that the sealant on the unformed web matches that on the formed web in order to achieve a proper seal. A variety of other compositions have been tried with varying success.

UNIWEB

Some bacon packers prefer to use a single material for both formed and unformed web. The supposed advantages are reduced inventory requirements, avoidance of using the wrong web on the wrong side, etc. One must, however, also consider the disadvantages of the uniweb approach. The total package cost may actually increase because the two sides of the package do not really require the same type of protective properties. Yet one must select the more expensive web as the uniweb in order to achieve the more stringent property requirements on the formed side. Also, Nylon is more heat sensitive than polyester and the application of heat seal bars to the surface of the Nylon may result in distortions, burnthroughs, wrinkles—all of which may increase the total failure rate.

FADE

This topic has been treated more extensively in reply to Q 6. With specific reference to the packaging of bacon it is sufficient to say that its meaty portion is subject to fade when exposed to UV light. Fade can be reduced appreciably through vacuum packaging and a UV barrier on the packaging material. A good vacuum packaging operation will deliver a bacon package with no less than 28 inches of vacuum. Such a package will survive exposure in a display case for a number of days without any visible fade. Thus a bacon package which is assured rapid turnover could well utilize a Nylon-olefinic homo- or copolymer material. However, during its life cycle a bacon package which is expected to have a longer shelf life will admit enough oxygen through the Nylon-polyethylene to reduce the vacuum level appreciably and at the same time admit enough UV light to cause serious discoloration of the meaty portion. By incorporating PVDC into the packaging material both oxygen and UV light are screened and the product survives a considerably longer period of time without fade (Figure 25-3). Many bacon packers have reached a compromise in which they utilize PVDC in their formed web where the PVDC screens both oxygen and UV and have omitted the PVDC in their unformed web where no UV exposure is encountered. This economy move, too, speaks against the above-named uniweb approach.

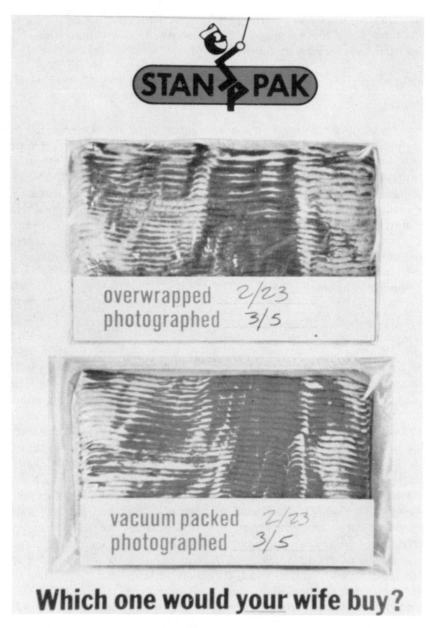

Figure 25–3.   Fade of packaged bacon.

CONVENIENCE FEATURES

Standard Packaging has in the late 1960s introduced a peelable as well as peelable/resealable patented packaging approach. The requirement for implements to open the package in order to get at its content was thus eliminated (Figure 25-4).

Separation of slices, compressed by vacuum and low temperature storage, has often been cited as a source of consumer annoyance. A patented process for interleaving strips of polyethylene and thus making the separation of slices a lot easier, has not been commercialized to this date. It has also been suggested that slice separation may be improved by gas rather than vacuum packaging. The equipment currently in use for vacuum packaging is easily adapted to gas flushing. Nevertheless, thus far this proposed solution has failed to gain commercial acceptance.

THE SLAB

The above discussion dealt primarily with the packaging of a shingled bacon product. The majority of shingled bacon is packaged in 1 lb quantity. However, there are also 8 oz., 12 oz., and 24 oz. packages on the market. In recent times a larger size vacuum package of bacon has made its appearance on the supermarket shelf. At least two packers are offering a sliced

Figure 25-4. The peelable bacon package. (*Courtesy Swift & Co.*)

slab of bacon in which the slices are restored to their original slab formation and packaged on a flat bed machine. The package contains a random weight of product in the range of approximately 1½ to 2½ lbs.

CASE STUDY 25-1

Swift research and development people were aware of a problem with their bacon package. In 1968, in their packaging laboratory at Oak Brook, Ill, they began looking at a new composite film that contained an inside layer of Du Pont's "Surlyn" ionomer resin in place of the polyethylene previously used. The head of the packaging research division, pointed out that when they first isolated the problem, they began looking for a material that would provide more packaging reliability. The problem was two-fold. First, there was the in-plant leaker problem where packages right off the machines were losing vacuum. Second, there was a need to improve the durability of the bacon package in the retail store.

Two features in the new composite film were required—a more reliable heat seal and a reduction in flex cracking. Several ionomer-based film constructions were tested by Swift. In use, the ionomer-based film selected reduced the number of leakers on their vacuum packaging lines. This was possible because of the improved heat seal characteristics of the ionomer-based film. Reliable heat seals even through bacon grease were no longer a problem on the packaging line. The improvement in flex crack resistance of the ionomer-based film in addition to the improved heat seal performance were important factors that contributed to a decrease in packages leaking at the retail store.

Supermarket introduction began in the Chicago area and soon extended into other cities in Illinois and in Indiana. To determine if the improved package performance observed by the Swift management was noticed at the retail store level, in-store audits of the leaker rates were conducted.

Over 22,000 packages were inspected for leakers in the Chicago area. This survey, in addition to the comments obtained by the meat department store managers, provided conclusive evidence that the in-store leaker rates dropped to a low figure which the store managers ranked equal to or better than any other vacuum bacon package on the market.

Upon completion of the Chicago area study, this ionomer-based packaging film was put into commercial use at Swift's Tampa, Fla., plant. Similar in-store audits of the packages were made in the Florida area after the introductory production period was completed. More than 14,000 packages were inspected in Florida. The in-store leaker rate reduction was confirmed by the audit.

Robert H. Lane, Production Control Manager for Swift Processed Meats

Company in Chicago, the man primarily responsible for the new package, reported that he considers the package performance excellent. The number of leakers attributable to the new composite film package was virtually zero, he noted.

ADDITIONAL READING

Anon. 1987. High quality package for fresh and processed meat products. *Food Engineering* **59**(1): 43.

Anon. 1976. The occurrence of *Clostridium* on vacuum packed bacon. *Journal of Food Technology* (UK) **11**(3): 229.

Anthony, S. 1988. Packaging quality = product quality. *Prepared Foods* **157**(3): 97–98.

Burner, R. 1988. Nylon for the nineties. *Plastic Technology* **34**(2): 60–65.

Ciani, L. A. 1972. Bacon buyers will see a slice. *Food Product Development* **6**(1): 34.

DeHoll, J. 1972. Get in on planning for new bacon packages. *The National Provisioner* **166**(23): 141.

Dempster, J. F. 1974. The effect of ultimate pH on some characteristics of vacuum packaged bacon. *Journal of Food Technology* (UK) **9**(2): 255.

Dempster, J. F. 1972. Vacuum packaged bacon: the effects of processing and storage temperature on shelf life. *Journal of Food Technology (UK)* **7**(3): 272.

Meeker, D. 1988. Atomic edibles. *Health* **20**(1): 65–68.

Post, L. S. et al. 1988. Development of staphylococcal toxin and sensory deterioration during storage of nitrogen and vacuum packed nitrite-free bacon-like product. *Journal of Food Science* **53**(2): 383–387.

Q 26: Which Is the Preferred Frank Package?

A 26: Frankfurters, frequently also referred to as wieners, are a mass produced processed meat item. The product differs widely in its composition. In all cases, it has a very high percentage of water content, meat, a relatively high fat content, and in some instances also cereal, spices, preservatives, and a variety of other additives. The meat may consist of one or a mixture of the following: beef, pork, and chicken. The ingredients are finely ground, masticated, and eventually emulsified. This emulsion is fed into casings. The frankfurters are thus cured and practically cooked while moving through a smokehouse. The modern smokehouse accommodates huge volume and delivers finished franks at such rapid rates as to require automatic handling of same at the pack off end.

"Peeled" frankfurters are fed automatically (see Figure 26–1) by an autoflex or similar unit onto a flatbed machine such as the Flex Vac 6–14. This equipment takes Nylon composites in roll stock form and after heating same, thermoforms a pocket into which the Auto Flex unit places frankfurters automatically. The most common configuration is one in which two layers of five franks each are placed into a pocket. The larger capacity of smokehouses and the need to pack franks at a more rapid rate has given rise to a second generation of automated machines in which a 50% increase in packaging capacity has been achieved. Frankfurters are oriented, collated, transported, and finally filled into the preformed pouch. During this process some breakage and some "skips" are normally encountered. The packer employs observers to remove broken franks and replace them with good product and to fill in product where same has been missed by the

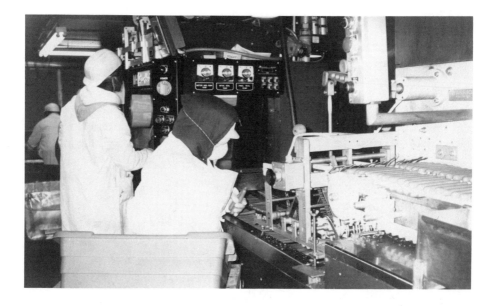

Figure 26-1. Automatic frank packaging. (*Courtesy National Foods Co.*)

automatic filler. A high degree of breakage is indicative of either a mechanical problem or a product deficiency. Insufficient cure or formulation errors should be considered.

It is important to move the franks rapidly into the package and offer the product vacuum protection. Contamination and concomitant microbial growth normally occurs at the surface. One must remember that in the frank the surface area per unit volume is almost eight times as large as that in a bologna chub. Thus the chance for spoilage is considerably higher.

THE PACKAGE

The most common frankfurter package consists of a thermoformed pocket in which ten frankfurters are placed. This formed material is transparent and normally consists of Nylon and a heat seal medium. Frankfurters are arranged in two rows of five each, stacked on top of one another. Occasionally, one sees larger franks with only eight to a package. One may also encounter two packages of five each, banded together.

The pocket is closed off with a "lidding" material. This material is in most instances printed. Sometimes the print is flexographic and very frequently it is gravure printing. The design is very important since it establishes the image that the packer wishes to convey to his customer. Examples of print designs are beyond the scope of this volume. However, the capabilities of gravure printing (especially in the case of Kroger, Ballpark franks, etc.) can and should be viewed in the supermarket. Figure 26–2 presents an example of a frank package.

Materials

There is no one preferred material in current usage. To minimize the large proliferation of different materials employed in the industry, one may recommend an inexpensive "economy," a "standard," and the expensive "deluxe" package.

Recommended Specifications

The formed web (FW) must be matched to the unformed web (UFW).

I. Economy	FW	0.001" Nylon/PVDC/0.002" PE
	UFW	0.0005" polyester/PVDC/0.002" PE
II. Standard	FW	0.001" Nylon/PVDC/0.003" PE
	UFW	0.0005" polyester/PVDC/0.002" PE
III. Deluxe	FW	0.001" Nylon/PVDC/0.002" ionomer
	UFW	0.0005" polyester/PVDC/0.002" ionomer

It should be noted that polyethylene will perform well in most instances. Where the packer feels that he needs improved performance he should change to the deluxe package, which includes a layer of ionomer.

Problems

Frank packaging is associated with a long list of problems. Breakage has been mentioned above and potential causes for the problem have been alluded to. Most other problems are associated with the quality of the product itself. A mechanical malfunction is rather easy to trace. Product spoilage due to packaging material failure is a very rare occurrence. More frequently one finds that the packaging equipment does not pull adequate vacuum and therefore the package appears to be loose. This can be avoided by periodic gauge checks to verify that the equipment does indeed produce the desired vacuum level. Seal bar damage can produce package failure. However, such failure would be repetitive in nature and should therefore be easily traced to its source.

A loose package could be indicative of many problems. Very often it is caused by "gassing." The product has either not been properly refrigerated or had an excessively high bacterial count and is thus decomposing even

Figure 26-2. Printed frank package. (*Courtesy Oscar Mayer*)

under vacuum. Decomposition generates gases which cannot escape the package and therefore cause ballooning. The package eventually will resemble a balloon. The product is certainly not fit for consumption.

Discoloration

Loss of color may be due to a variety of causes. UV light may contribute to the discoloration of franks. The presence of PVDC in the packaging material aids in reducing the effect of UV light on the meat. It is thus essential to incorporate PVDC into packaging materials for franks, in order to reduce the admission of oxygen into the package and thus maintain a higher level of vacuum for a longer period of time and also because of its UV light screening properties. However, the discoloration may also be due to the action of bacteria or mold. The first type of discoloration discussed is not harmful and merely diminishes the esthetic quality of the frank; the latter may indeed be indicative of poor sanitation.

Slime or a milky white fluid is often noticed in frank packages, especially those of somewhat advanced age. The frank does contain a high percentage of water. When packed under vacuum, pressure is placed on the frank which squeezes the product and forces some of the water out into the package cavity. Lactic acid bacillus, which by the way is quite harmless, will grow under normal storage conditions in this fluid. Lactic acid bacillus is facultative in nature; this means that it can grow in air or in a vacuum. It also has the ability to multiply at relatively low temperatures. The growth of this bacillus presents no particular danger—quite the contrary, it acts as a potential safety valve. It has been determined that in the presence of lactic acid bacillus the growth of botulinus is retarded.

Thus a potentially hazardous germ, which could also grow under a vacuum, is prevented from flourishing by the presence of these harmless bacilli. Many a consumer will find the appearance factor of slime generated by lactic acid bacillus objectionable. From a purely safety point of view, all that needs to be done is to wash the franks prior to heating them.

CASE STUDY 26-1

Any one of the packages recommended above is not meant for freeze-thawing. It is quite all right to use any package for home freezing. The housewife uses the package she picks up at the supermarket and places same in her freezer compartment. Product will survive for an almost indefinite period of time under frozen conditions. However, if product is frozen by the packer and shipped in a frozen state and thereafter thawed prior to retail sale, then there are usually a great many problems associated with this. The difficulty is not just one associated with the packaging material. It is inherent in the product itself. Because of the high water content, freezing forces some of the moisture into the empty spaces inside the package and it freezes like icicles. Shipments with sharp pieces of ice in a relatively thin flexible package cause a high failure rate. Furthermore, upon defrosting the product just does not look presentable anymore. It is very likely that the product would have to be reformulated for freeze-thaw distribution.

ADDITIONAL READING

Anon. 1987. Nitrate, nitrite and nitroso compounds in foods. *Food Technology* **41**(4): 127–134

Anon. 1987. High quality package for fresh and processed meat product. *Food Engineering* **59**(1): 43.

Hirsch, A. 1973. Sausage package—selling the sizzle. *Gravure* **19**(8): 111.

Kelsey, R. J. 1988. Robots in packaging. *Food and Drug Packaging* **52**(3): 45–47.

Pietraszek, G. 1988. Canadian meat scientists focus on MAP consideration. *The National Provisioner* **198**(11): 9–13.

Q 27: Which Is the Best Luncheon Meat Package?

A 27: A large variety of processed meat products are sold under the collective name of "luncheon meat." A good portion of luncheon meats are sold prepackaged in the supermarket. This is not meant to deprecate the more traditional sale at the local deli or the growing activity at the supermarket deli counter. The consumer very often prefers to have bologna sliced specifically for his (or her) demand. It must be realized that both modes of marketing will survive for some time and actually compliment one another.

The prepackaged luncheon meat fits the modern system of retailing. If pickles, ketchup and mustard are readily available to the consumer on the open shelf, why not the processed meat as well? Both products and package have severe traditional limitations imposed on them. Bologna is by tradition round, and no one would dare make a square bologna, for example. Thus one is constricted by past usage to a limited number of shapes and sizes.

The 8 oz. package seems most popular. Recent attempts at introducing 2 oz. packages geared to the needs of the "singles" trade has met with little success. At the other end of the size spectrum, we find a limited market for the 16 oz. package. Other sizes such as 4, 6, and 12 oz. enjoy a portion of the existing market.

THE MACHINE

In spite of the large variety of packages in current use, there are certain common features applicable to all luncheon meat packages. They are all vacuum packages formed on a flatbed machine. The most popular machines include Mahaffy-Harder, Royal Vac, and Multi Vac. Not all of the aforementioned are capable of producing all of the packages described herein.

Each machine has its limitations and represents a compromise between cost and various capabilities. The meat packer must decide, on the basis of his own priorities, the selection which best meets his specific needs.

Some limited quantity of luncheon meat is still provided in a premade pouch (as distinct from roll stock). This is a mode of packaging utilized by packers of limited volume or specialty products. Pouches were traditionally cellophane/polyethylene. The cellophane provided a glasslike appearance, an excellently printable surface, and a degree of stiffness most desirable for shelf display. Normally the pouched luncheon meat was packaged under gas. There is a trend away from the pouch to roll stock. Those who continue in pouches are giving serious thought to plastics, such as polyester, Nylon, and polypropylene to replace the costlier cellophane. However, ecology considerations have given cellophane a new chance (see Q 24).

THE ALL-FLEXIBLE PACK

The least expensive roll stock package is produced from a composite consisting of nylon/polyethylene as a thermoformable web and polyester/polyethylene as an unformed web. Gauges employed depend on depth of draw (4 oz. versus 16 oz.), type of product (with or without casing, for example), shipping radius, etc. For especially difficult circumstances substitution of ionomer as sealant is considered by some to be beneficial.

SEMI-RIGID

The concept of placing luncheon meat into a transparent rigid tray was originated by Oscar Mayer and has since been imitated by every major national and regional meat packer. Formerly, PVC with or without sealants (PE, EVA, or ionomer) was utilized. In the mid-1970s the FDA threatened to ban the use of PVC and a flurry of activity to replace same ensued. At this writing, several plastics are jockeying for position with no clear-cut favorite emerging (Table 27-1).

The lidding material, in most instances, is colored. The preferred color is yellow, but orange, red, and other colors are in use as well. The composite is 50 ga. oriented polyester with 2 or 3 mils of a sealant. A formable polyester is employed for machines that have back forming capabilities. In such instances, the rigid tray has a slightly larger than required depth. The lidding material is drawn into the void upon evacuation, giving the package a countersunk appearance.

The semi-rigid package is normally provided with a header. A center hole

Table 27-1. Properties of Converted Semi-Rigids.

Test	0.0075″ PVC/ 0.0015 PE	0.0075″ PET/ 0.0015 PE	0.0075″ CoPET/ 0.0015 PE
Tensile (psi)	45.3	45.0	63.9
Elongation (%)	77	376	319
Impact (dart w-50, g)	851	1000	1000
Tear (Elmendorf, g)	512	517	709
Mullen burst (psi)	105	100	126

permits pegboard display of the product. This is a major marketing advantage, moving the product from the bottom of the refrigerator case to the consumer's eye level. The header is also utilized as a display panel for label attachment. In the all-flexible package, the label is normally centered on one of the major display surfaces, blocking the view of a good portion of the product. In a semi-rigid package the obstruction to the product display has been moved to the header.

CONVENIENCE FEATURES

The advent of semi-rigid packaging permitted the introduction of several convenience features to meet customer needs. One of the primary advantages of the rigid tray is its potential utilization as a serving tray.

The consumer's difficulty of opening the package has been alleviated by an "easy opening" mechanism. A further improvement, patented by Spiegel et al. (Figure 27-1) is an easy opening/resealable package. First introduced by Swift & Co., this slightly more costly package provides the added convenience of converting the package into a storage tray for leftovers.

Recommended packaging materials are found in Table 27-2. There are many variations in use, all designed to meet the special needs of the individual packer. The rigids shown in Table 27-2 could be PVC, PET, CoPET, Barex, or others.

FADE

The fade of luncheon meat, bologna specifically, has been studied. The utility of PVDC in the packaging material to prevent or at least minimize this undesirable phenomenon has been documented. The mechanism through which PVDC provides this protection, however, seems debatable. One theory has it that PVDC is not a light screen, but rather aids in achieving a tight cling of packaging material to the product surface and thus reduces surface evaporation. However, no matter what the mechanism, the

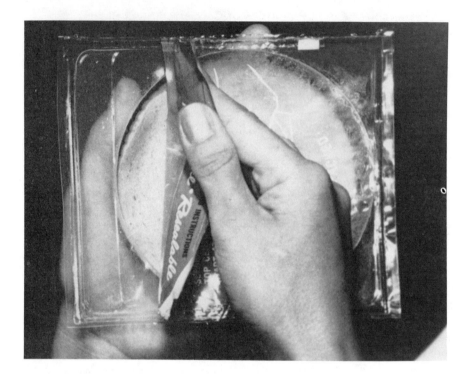

Figure 27-1.   Peelable/resealable luncheon meat package.

Table 27-2.   Luncheon Meat, Recommended Film Composites.

Flexible Package	
Formed	1 mil* Nylon/PVDC/3 mil LDPE
Unformed	0.5 mil PET/PVDC/2 mil LDPE

Semi-Rigid Package	
Formed	6 mil rigid/1 mil LDPE
	7.5 mil rigid/1 mil LDPE
	10 mil rigid/1.5 mil LDPE
Unformed	0.5 mil PET/3 mil LDPE

Peelable Ionomer Package	
Formed	rigid/2 mil ionomer
Unformed	0.5 PET/PVDC/2 mil ionomer/heat seal coating

*1 mil = .001 inch.

fact remains that "PVDC reduces fade in vacuum packaged processed meat."

CASE STUDY 27-1

A nationally (or even internationally) well known meat packer recorded a high failure rate in his processed luncheon meat operation. He blamed his packaging material supplier, the machinery manufacturer, the packaging materials, etc. At no time was he willing to face up to the real cause—his own operation.

This packer was taken on a tour of a competitive operation employing the very same equipment, the same packaging materials furnished by the same supplier. Yet this good regional packer had less than ½% rejects, while the national operation reported close to 7% failure. An audit of the national packer's operation netted a long list of recommended changes in operating procedures. On the top of that list should have been "management attitude." A followup audit, six months later, revealed that none of the changes had been instituted. Management had decided to seek easy solutions and had instead forfeited its position in the luncheon meat market.

CASE STUDY 27-2. OVERTREATMENT OF FILMS

Electrostatic corona treatment is widely used to upgrade the surface energy of plastic films, which in turn improves wettability and adhesion of these otherwise nonpolar surfaces. Response to treatment varies according to polymer composition and condition. Freshly made film generally is easier to treat than stock that has been in shipment and storage over weeks or months. Unfortunately, treatment level often decreases with time. This necessitates retreatment in many instances.

A number of converters of PET film encountered an interesting set of symptoms, all within a period of months. If these incidents were spread over a longer period, attention might have been diluted. Treatment at the unwind stand, regardless of incoming status of the film, is standard procedure in some facilities. At times, buildup of brown particle residues was noted on applicator rolls, idler rolls, primer pans, and other equipment. Investigation disclosed a relationship between presence of the buildup and strong or excessive surface treatment. The basis for the particles is the presence of PET oligomers, or low-molecular-weight portions in the PET film. They were formed in the course of extruding and orienting the film. Additional quantities of oligomers can be formed by excessive treatment or retreatment. Sufficient oligomer content can form a weak boundary layer resulting in low bond level or even bond failure when producing a laminate!

If insubstantial bonds formed in this manner succeed in passing quality checks, subsequent failure can occur on the packaging line, in the refrigerated showcase, or in some other stressful condition that normally would be easily tolerated.

At one converter, this experience caused a change in procedure resulting in treatment of PET to a surface energy level only slightly higher than the minimum required for adhesion. The previous and prevalent attitude was "the higher, the better." If PET film is received with a surface treated by the manufacturer, these converters generally no longer retreat.

While this potential problem is not peculiar to extrusion coating, the presence of particle buildup appears to be more apparent in the environment of aqueous primers than with solvent-based adhesives.

ADDITIONAL READING

Andres, C. 1976. Sliced luncheon meats are bulk packed. *Food Processing* **37**(4): 92.

Anon. 1976. Labeler, multi-cavity line speed bologna packaging. *Food and Drug Packaging* **34**(8): 3–8.

Anon. 1976. Individual portion packs get test marketing in south. *The National Provisioner* **24**(18): 25.

Anon. 1976. Nitriles moving past the pop bottle. *Package Engineering* **21**(9): 38.

Hirsch, A. and Spiegel, F. X. 1975. PVDC prevents fade of processed meats. *TAPPI PVDC/Adhesives Laminating Seminar.* Hilton Head Island 11/18/75.

Monia, V. and Spiegel, F. X. 1969. Easy opening and reclosable package. U.S. Patent 3,423,762.

Q 28: How Can We Package Fresh Meat?

A 28: Fresh meat—beef, veal, pork, and lamb—marketing has undergone some major changes. The small butcher has practically disappeared. Most meat is sold in the supermarket. However, the supermarkets have failed to modernize the butchering operation sufficiently to keep pace with advances in technology. Some claim that the high cost of meat is the consequence of this outdated processing and marketing technique.

In order to reduce, or preferably eliminate, the time and space consuming operation at the retail level, it is mandatory to impart longer shelf life to individual meat cuts offered to the consumer. Extending shelf life would open the door to regional or possibly even national distribution of retail meat portions.

In addition to reducing overhead, minimizing spoilage, and ensuring better quality control, since meats could then be checked and processed from a central source, it might even enable meat prices to stabilize at lower levels. Without question, improved packaging could transform the outmoded butcher shop operation into a modern, more profitable 21st Century retail meat section.

FRESH VERSUS FROZEN

With nearly every other type of foodstuff being retailed in a frozen state, it seems logical that meats could be sold frozen. However, many years of extensive test marketing have shown that the average consumer is not prepared to purchase quality frozen meat. While the consumer may purchase

177

fresh meat, take it home, and freeze it, this same buyer rejects the fresher quick-frozen meat at the butcher counter.

The question is, why? American consumers are most progressive. In the last 30 years, they have seen and readily accepted major innovations in food packaging. Consumers once upon a time bought fresh vegetables only. They now accept canned and frozen produce readily. In addition, they have eagerly accepted instant coffee along with countless precooked, dehydrated, freeze-dried, and other ready-to-serve products and have even shown a willingness to pay a premium for them.

The consumer's rejection of frozen meats obviously complicated matters for the distributor as well as the packager. In a frozen state, meat could be stored for many months and shipped over long distances, making central meat packing an easily attained reality. Bloom, the bright red color of the meat, could be retained throughout storage in a properly selected packaging material. All this has been available for sometime (see Bivac below), but the public must be educated to accept this packaging concept.

A misconception of the consumer is the belief that "red meat is fresh meat," requiring present and future distribution systems to deliver red meat in full bloom to the consumer. Even institutional buyers require this color conformance.

NEWER PACKAGING METHODS

One of the innovations of the 1960s and 1970s was the "subprimal" package. This method involves the packaging of larger section of the carcass in larger plastic pouches. This method provides a partial vacuum to prolong storage life. Although this type of packaging does not retard enzymatic aging, it does offer several advantages, among them permitting shipment of meats from a central source.

It also has several major drawbacks, since additional operations at the retail level reduce the meat's potential shelf life, increase the product's cost, place an additional burden on packaging personnel, require handling space in the cutting area and prevent maximum utilization of trim and bones.

Bivac

This system, developed in the late 1970s by Bill Young and financed by E. I. Dupont, was designed to package retail cuts on a relatively inexpensive machine, between two layers of ionomer. The film provided strength and enough oxygen permeability to retain the bloom. The package was designed to sell fresh frozen meat. But it could be utilized for fresh sale as well. At refrigerated temperatures, the ionomer would reduce the need for refacing due to moisture loss, but would not increase the shelf life due to bacterial growth. This system, while very attractive in many respects, has not made any major inroads in the market place.

Vacuum Pack

Red meat turns purplish under vacuum. Consumers are reluctant to accept purple meat, just as they decline the frozen meat option. Once the vacuum package is opened, the meat will bloom again. However, the consumer is guided by the first impression only.

Controlled Atmosphere

In the late 1970s and again in the 1980s a novel system for central meat packaging was test marketed. The meat would be cut to retail (or restaurant) portions and packaged ready for store display. Since the package would have appreciable shelf life, there would be no further need for instore butchering activity. The packaged portion, with some judicious handling, required no further skilled butcher to reach the ultimate consumer. Many studies have been undertaken to prove that the stores, the institutions, and the consumer would be better served by such a packaging system.

Several false starts included American Beef, Iowa Beef, Safeway, Goehering Meat, and others. The industry lacked commitment and wanted to avoid facing up to union objections. The central meat cutting system would no doubt displace many butchers from the supermarkets. However, new jobs at central locations would be generated to offer new opportunities to less skilled meat cutters. Besides,the highly specialized trade of butchers was rapidly becoming extinct and the new system offered a solution to this problem.

In the 1980s there were several other attempts to adopt a CA package for retail meat packaging. Kroger Co., one of the largest supermarket chains in the U.S., tested the "super fresh system" developed by St. Regis. This system was in essence the same as the Standard Packaging approach detailed above. Another similar method was tried in Canada. But none have caught on.

The system was designed both to extend the shelf life of fresh meat products without freezing and to provide a method of preparing a retail package centrally.

This package involves specially designed oxygen-impermeable packaging materials, including a multilayer lid stock, a rigid forming material and a packaging machine which allows for preparation of the grooved trays, and the development of an inert environment interior to the package, all this yielding the hermetically sealed retail package.

During the sealing operation, the package is first evacuated, then a carefully selected inert gas mixture of CO_2/N_2 is bled back into the system. With the resultant low residual oxygen level developed (below 0.3%), the hermetically sealed meat products' quality is maintained for a considerably longer period of time than could be attained with the current overwrap display technique.

At the retail store level, just prior to placing the package on display, the lid is peeled off. During this peeling process, the lid separates, allowing for the removal of the oxygen impermeable layer but leaving behind a permeable membrane film covering the tray. This layer remains sealed to the tray, thus insuring a hermetically sealed package at time of display and sale.

This membrane layer allows the air with its oxygen to enter into the package, thus regenerating meat product bloom but preventing contamination of the packaged meat. In addition, moisture loss of product that is displayed in this manner has been demonstrated to be virtually nonexistent.

In spite of all the advantages enumerated above and the scientific evidence of viability demonstrated (see Case Study 28–1), the modern meat packing continues to be a good idea who's time has not yet come.

CASE STUDY 28–1

A regional meat packer on the West Coast wanted to evaluate the central meat packing concept. Since extended shelf life is best demonstrated by reduced bacterial activity, a pork product was selected for microbiological examination; pork has a "difficult to preserve" reputation.

Pork chops, steaks, and various other cuts were packaged on a Flex-Vac Mark X Rollstock packaging machine. This equipment was outfitted with specially designed die cavity inserts which produced a thermoformed tray of a configuration quite similar to those in wide use in supermarkets today. Tray sizes were 4-15/16 × 5-1/2 in. or 4-15/16 × 12-1/8 in., and depths ranged from 1 in. to 2-7/8 in., inside dimensions. A wide range of other size trays can be prepared on several other Flex-Vac machines as well.

The packaged product selected for the test was maintained at 36–38°F

Table 28–1.   Color Regeneration.

Storage Time	Air	Vacuum	Gas
14 days	Fair	OK	OK
22 days	Poor	OK	OK
38 days	—	OK	OK

Table 28-2. Odor Generated.

Storage Time	Air	Vacuum	Gas
14 days	Slight	OK	OK
25 days	—	OK	OK
38 days	—	Slight	Slight

and was sampled twice a week for olfactory, visual, and bacteriological analysis.

Bacterial testing involved the sampling of one square inch of meat surface, blending with sterile dilution water, and taking plate counts using TGEE agar, ABA base, McConkey's agar, and *Lactobacillus* agar.

Each testing cycle also involved package preparation for retail display (membraned state), and the packages were reexamined at intervals for up to eight days.

For comparison purposes, sample units studied were air, vacuum, and the preferred inert gas-filled packages.

Peeled packages were assessed both as to return and retention of bloom. As can be noted in Table 28-1 color regeneration was determined to be good in both vacuum and gas packages even after 38 days of storage, while the air packs lost color in less than 14 days.

With full bloom being returned in approximately 15 to 45 minutes after barrier layer removal, the meat retained its color for up to seven days in retail display.

The color of the gas-packaged product proved to be superior, for reasons not established as yet.

USDA Choice beef, similarly packaged, including a variety of cuts, yielded results that were parallel to those outlined above.

The results of odor determination are summarized in Table 28-2. While

Table 28-3. Bacterial Studies, Barrier Retained Condition, Total Count, Microorganisms/Sq. In. Meat Surface.

Storage Time	Air	Vacuum	Gas
0 days	8.0×10^2	1.4×10^3	9.0×10^2
7 days	6.0×10^4	3.0×10^3	4.0×10^2
14 days	3.0×10^8	4.4×10^4	—
22 days	T.N.T.C.*	9.0×10^4	7.4×10^4
28 days	—	1.7×10^6	3.2×10^7
35 days	—	5.3×10^7	2.0×10^7
42 days	—	1.1×10^9	1.0×10^9

*T.N.T.C. denotes "too numerous to count" at $1 : 10^8$ dilution.

Table 28–4. Bacterial Studies, Barrier Retained Condition, Gram Negative Count, Microorganisms/Sq. In. Meat Surface.

Storage Time	Air	Vacuum	Gas
0 days	2×10^2	LT*	LT
7 days	1×10^2	LT	LT
14 days	2.5×10^8	1.0×10^3	—
22 days	T.N.T.C.*	8.0×10^3	1.4×10^4
28 days	—	2.0×10^4	1.5×10^4
35 days	—	6.0×10^5	4.8×10^5
42 days	—	2.4×10^9	7.8×10^8

*T.N.T.C. denotes "too numerous to count" at 1 : 10^8 dilution.
LT denotes less than 100.

the air packs evinced surprisingly low odor development, even up to 14 days, due probably to extremely low initial bacterial counts, both the vacuum and gas packages exhibited acceptable odor levels up to 38 days. This odor did not become what was described as disagreeable until after 45 days.

Tables 28–3 and 28–4 summarize both total count and gram negative (spoilage) bacteria development results in the sealed package.

The determinations clearly show the comparatively slow growth rate in both the vacuum and gas systems, particularly for the first three weeks.

The control air packs, on the other hand, reached an intolerable 10^8 count in less than 14 days, as would be expected.

Due possibly to a reduced pH level as a result of CO_2 inclusion in the packaging cycle, the gas packages consistently yielded lower counts than did the vacuum pack equivalent. McConkey's agar was used to monitor gram negative bacteria development, principally *Pseudonomas* and *Achromobacter*. These results are summarized in Table 28–5.

Tryptone glucose yeast extract agar was used to obtain total count deter-

Table 28–5. Bacterial Studies, Retail Display State, Gram Negative Count, Microorganisms/Sq. In. Meat Surface.

Days of Storage	Days Membraned	Package Environment	
		Vacuum	Gas
0	3	LT*	LT
3	4	LT	LT
7	3	2.9×10^3	2.0×10^2
14	3	5.8×10^3	6.8×10^3
22	3	2.1×10^4	2.5×10^5
25	3	6.0×10^4	6.5×10^3
35	3	4.0×10^6	2.7×10^6

*LT denotes less than 100.

Table 28-6. Bacterial Studies, Retail Display State, Total Count, Microorganisms/Sq. In. Meat Surface.

Days of Storage	Days Membraned	Package Environment	
		Vacuum	Gas
0	3	4.3×10^3	4.1×10^3
3	4	2.0×10^2	2.7×10^3
7	3	1.4×10^4	7.0×10^2
14	3	4.8×10^5	7.0×10^4
22	3	4.0×10^4	4.0×10^4
25	3	8.0×10^4	4.2×10^5
35	3	1.6×10^8	3.0×10^9

minations. The same pattern emerges from these results summarized in Table 28-6.

The system unquestionably works. The standard supermarket pork chop loses 100 g of moisture each day—the CA package loses no more than 0.98 g/day. A truck shipping test over a 4000 mile distance produced 3 out of 200 leakers (1.5% failures). Better boxing and strapped palletizing could reduce this appreciably. A 45-day shelf life for pork is a very significant achievement.

ADDITIONAL READING

Anon. 1987. High quality packing for fresh and processed meat products. *Food Engineering* **59**(1): 43.

Anon. 1984. Controlled atmosphere extends meat shelf life. *Packaging* **29**(8): 8.

Hirsch, A. et al. 1977. Controlled atmosphere packaging. *U.S. Patent* 4,055,672.

Hirsch, A. 1975. Packaging technique shows promise for central cutting. *The National Provisioner* **185**(5): 24+.

Hirsch, A. 1975. A new approach to fresh meat packaging is inevitable. *Meat Processing* **14**(6): 46+.

Lee, B. 1988. Nitrogen vs. vacuum package as affecting meat spoilage. *Food Processing* **49**(2): 136-137.

Lee, J. Y. and Fung D. Y. C. 1986. Method for sampling meat surfaces. *Journal of Environmental Health* **48**(1): 200-205.

Morris, C. E. 1988. Vac-pac CAP spur case-ready meats. *Food Engineering* **60**(7): 73.

Rothwell, T. T. 1988. A review of current European markets for CA/MAP. *Food Engineering* **60**(2): 36-38.

Schillinger, V. and Lucke, F. K. 1987. Lactic acid bacteria on vacuum-packed meat and their influence on shelf life *Die Fleischwirtschaft* **67**(10): 1244-1248.

Tharrington, J. 1988. Vacuum in the meat case. *Meat and Poultry* **34**(1): 164-167.

Q 29: What Is the Preferred Cheese Package?

A 29: There are several hundred types of cheeses on the market. This discussion will be limited to hard cheeses. For clarification, soft cheese examples are cottage, cream, and farmer cheeses. The hard cheeses include: swiss, munster, american, gouda, etc.

Most prepackaged hard cheese is sold in either slices or chunks. Both of these packages are made on a Hayssen or Campbell wrap machine. This mode of packaging has replaced the earlier pouch package and is in turn experiencing some pressure from a newer thermoforming vacuum package.

The Hayssen machine produces a pillow pack. To protect the product from molding, oxygen should be excluded. Normally, carbon dioxide is employed to sweep the oxygen from the package. The residual CO_2 exerting a positive pressure converts the package into a balloon having the appearance of a pillow. The air is not fully replaced, leaving up to 5% oxygen in the pillow pack. This naturally limits the shelf life of the product.

The pillow pack has a further disadvantage of requiring excessive quality control. The packages coming off the line are indistinguishable from air packs. It is therefore necessary to analyze the gas content of packages at frequent intervals to assure that the carbon dioxide fill is indeed taking place. As the package ages, the CO_2 is absorbed into the cheese and the packaging material is drawn tightly around the product.

The pillow pack requires excessive quantities of packaging material, as may be seen in Table 29-1. The quantity of material consumed on a Hayssen machine may be more than double that needed for a vacuum pack.

THE MATERIALS

The most widely utilized composite in pillow packs is a triple-ply material consisting of OPP/cellophane/EVA. Originally, cheese was packaged in

184

Table 29-1. Packaging Material Requirements, Cheese Packaging.

	Square Inches/Package	
Package Size	Pillow Pack	Vacuum Pack
8 oz	73.5	32.48
12 oz.	80.75	45.14
16 oz.	90.1	55.18

cellophane/polyethylene pouches. When the pillow pack was introduced, it was found that the two-ply material failed due to flex cracking. The excess material at both ends of the package undergoes considerable flexing during packing and while on display at the retail store. Furthermore, because the pillow pack requires appreciably more space at time of cartoning, prior to absorption of the gas, it is necessary to utilize oversized shipping cartons. During transit the excess space generated by CO_2 absorption permits movement of the packages with concomitant excess flexing.

To overcome the flex cracking problem and provide improved cold weather durability, a third layer was superimposed on the cellophane/polyethylene composite. The outer ply consists of oriented polypropylene. Since the heavier composite exhibits poorer thermal transmission it became necessary to substitute a sealant that can be activated at lower temperatures. EVA was thus chosen.

In recent times several alternative film composites have been introduced for the purpose of reducing cost and/or improving performance. For less demanding package performance one notices increased use of a two-ply sheet. This material is composed of a somewhat heavier layer of polypropylene and EVA. Nevertheless, it lacks the stiffness of the three-ply sheet and does not machine as well as the standard material.

Because of the increased cost of cellophane and also due to inherent weather related problems of this sheet, substitution for same in a three-ply structure has improved the durability of the package. Both polyester and nylon have been used with good success.

VACUUM PACK

The vacuum package offers distinct advantages:

Speed

Vacuum packaging machines can run up to twice as fast as gas pillow pack equipment. Thus for the same floor space and manpower one can generate double the number of packages.

Material Savings

As mentioned earlier, there are appreciable material savings possible when utilizing a thermoforming vacuum machine.

Failure Rate

There are no excess materials in a vacuum package. The packaging material clings tightly to the product as it comes off the machine and flexcracking is thus minimized if not totally eliminated.

Quality Control

Concern for frequent QC sampling is obviated. The quality of the finished package can be visually ascertained almost immediately off the packing line. Delayed gas absorption has been eliminated.

Even though the vacuum package has all these and many other assets to recommend it, it has been slow in gaining market acceptance. It must be assumed that a primary reason for this delay in switching to vacuum packaging of cheese is the need for capital investment. Most cheese packers are heavily invested in the older equipment and will not switch until they have to purchase new machinery. One can observe this phenomenon at several cheese packers who have added to their several gas lines a new vacuum type of equipment. In most instances this is a dieless machine of the Multi Vac or related class. Some of these machines are capable of packaging vacuum or gas and thus give the cheese packer an option to produce packages of either type.

SOFT CHEESES

Most of the discussion dealt with the packaging of hard cheeses such as american, swiss, muenster, etc. An appreciable volume of varied soft cheeses is sold in the U. S., as well. A considerable portion, such as cream cheese, cottage cheese, and others, are sold in rigid tubs. Some cream cheese is found in foil overwrap. This product is sold in sticks or squares, protected by a simple printed foil laminate. Some mozarella cheese can be found in a vacuum package.

ADDITIONAL READING

Anon. 1987. Zipper reseals cheese packages. *Packaging* 32(3): 51.
Anon. 1986. Tub replaces crock. *Packaging* 31(1): 11.

Anon. 1985. Unpretentious convenience. *Packaging* **30**(2): 10.

Anon. 1973. New packaging methods increases cheese productivity, reduce manufacturing cost. *Modern Dairy* (Canada) **52**(6): 20–22.

Anon. 1972. Vacuum packers extend life of blocked cheese. *Food and Drug Packaging* **27**(5): 1.

Hartmann, G. 1972. On the behavior of cheese portions in the laminated film packages. *Verpackungs Rundschau* **23**(6): 732–734.

Kiermeier, F. et al. 1972. On the behavior of cheese in the package. *Zeitschrift fuer Lebensmittel-Untersuchung und Forschung* 149(3): 156–166.

Knill, B. J. 1987. Cheese pack reseals via zipper. *Food and Drug Packaging* **51**(2): 3, 21.

Nelson, K. M. et al. 1973. Bringing and packaging of cheese. *Dairy and Ice Cream Field* **156**(6): 36–41.

Scott, R. 1972. Cheese packaging. *Journal of the Society of the Dairy Industry* (UK) **25**(2): 98–101.

Taylor, D. L. 1985. Tougher pasteurization laws ahead for the California dairy industry. *Food Engineering* **57**(8): 25.

Q 30: What Choices Are Available for the Packaging of Coffee?

A 30: The packaging of coffee presents a large array of challenges, opportunities, and difficulties. Coffee gave the impetus to modern packaging and especially to controlled atmosphere (CA) packaging. Back in the 1950s an Italian machine was imported to vacuum pack coffee. Unfortunately, the idea, while very laudable, never worked. The vacuum lines would fill up with ground coffee before the package could be sealed. Eventually, the coffee project was abandoned, but the experience gained helped to found a new industry, CA food packaging. The engineers on this commercial failure went forth and founded a number of successful companies, many of these are still in the forefront of new package design and development.

CONSUMPTION

The consumption of coffee has been on a downward trend. In 1962 per capita consumption amounted to 3.12 cups per day. In 1989 this diminished to 1.75 cups per day. Also the ratio of people drinking coffee has declined from 74.7 to 52.5 percent of the population.

The per capita consumption of decaffeinated coffee between 1962 and 1989 has increased from 0.10 to 0.40 cups/person/day. However, even in 1989 this is less than 12% of total consumption. Most of the coffee is of the regular grind, while just a little more than 22% is instant.

Coffee remains the second most popular beverage in the U.S.—more than 52% of the population drinking it. (Number one in popularity are soft drinks). Milk and juice are not very far behind in market share. There seem

188

to be some regional differences—with North Central showing the highest and South the lowest consumption. This could be possibly climate related.

THE MARKET

Very little coffee is imported in a processed form. Almost all of it comes into the U.S. from South American plantations as raw beans and is processed by:

- National roasters, such as General Foods, Procter & Gamble, Hills Bros., or others.
- Trade roasters, who supply small chains, specialty stores, some restaurants, etc.
- Regional roasters, who manufacture brands distributed in chains and independent supermarkets—such as Savarin, Chock-full-o-Nuts, etc.
- Roaster-wholesalers, who carry their own label.
- Roaster-retailers, which are large chains, such as A & P or Safeway, that run their own roasting operation.

Three or four brands dominate the retail market, with all others accounting for less than 45% of the U.S. usage. In the institutional market there is no such clear-cut dominance. It is very likely that local or regional roasters will dominate in limited areas due to better service of accounts in the vicinity of their operations.

Most coffee is consumed in the home and most of it is packaged in cans or glass jars. Almost 40% of the coffee is sold into the institutional market. This last category is somewhat misleading, since it implies hospitals, old age homes, etc. Instead, the majority goes into vending machines and close to 40% of this market consists of hotels and restaurants.

THE PACKAGE

Ground coffee is very sensitive to oxygen due to its oil content. Because of its high surface area, the ground coffee will stale quickly, lose flavor and aroma, and become rancid.

The selection of package must take into consideration the distribution and consequent shelf life expectancy. The national distributor places the ground coffee in a can under vacuum or nitrogen and can count on a 6 month shelf life. The product that is ground in the store is placed in a lacquer-coated paper bag.

INSTITUTIONAL

Hardly any coffee for this market is in cans. Some is still in paper bags. For cost consideration, however, there has been a large shift to metallized polyester or polypropylene bags. The metallized material is more expensive, but could in the long run be the more economical choice.

The packaging requirements for institutional coffee are:

1. Good durability.
2. Shelf life, depending on distribution range.
3. MVTR (less than 0.3 g/100 sq. in./24 hrs).
4. Good O_2 barrier.
5. Impermeability to odor and slight to CO_2.
6. Good machinability for ffs operation.
7. Acceptable leaker rate.
8. Low cost.

The more expensive metallized package may prove less costly in the long run, since it improves all of the above requirements. Shelf life especially can be improved very appreciably, thus cutting distribution cost. Vendor-salespersons are free to make fewer deliveries. Production in the roasting plant can work from stock, rather than from machine to delivery wagon.

Leaker rate has been drastically cut with metallized films. Other flexible materials would pick up static electric charges as they moved over the filling equipment. The coffee grinds would cling to the inner face of the packaging material, lodging in the seal area and causing imperfections in the seal. In a gas or vacuum package this type of annoyance often resulted in leakers. The metallized film has antistatic properties and this practically eliminates leakers from this source. The savings realized from shelf life extension and leaker reduction pay for the increased package cost. Some of the advantages and shortcomings of both the gas flush and the vacuum package are summarized below:

Vacuum Packaging versus Gas Flushing for Institutional Coffee.

Vacuum Packaging	
Advantages	Disadvantages
1. Extend shelf life.	1. Not suitable for sharp particles.
2. Condensation minimized.	2. External abuse creates strain on packaging material with subsequent high leaker rates in field.
3. Weight loss prevented.	3. Product can be crushed.
4. Quick check for leakers.	4. Poor appearance, wrinkled and distorted.
	5. Equipment very costly and slow production rates.

Gas Flush	
Advantages	Disadvantages
1. Extended shelf life.	1. Inert gases, i.e., nitrogen will not retard product deterioration after initiated.
2. Provides low oxygen level *without* high internal vacuum.	2. Leakers more difficult to locate.
3. Will not crush product.	3. Longer holding period required to check on seal performance.
4. Much better customer appeal from appearance standpoint.	
5. Equipment much less expensive and higher production rates obtainable versus vacuum pack.	
6. Less durable film structure acceptable and functional at low cost.	

An Interesting Comment

The consumption of coffee has been declining for a variety of reasons. Caffeine is one of the excuses often cited. A study reported in the summer 1990 issue of the *Quarterly* (University of Wisconsin, College of Agriculture and Life Sciences) may shed light on benefits derived from caffeine. Lewis Sheffield reported that pregnant mice getting daily doses of caffeine produced more milk and raised heavier offspring than caffeine-free mice.

MACHINERY

Some of the more widely used equipment in coffee packaging is offered as a source for further inquiry. It is by no means all inclusive.

Gas Flush Coffee Packaging Machinery Suppliers.

Machinery Manufacturers	Location	Telephone Numbers
FMC Corporation Packaging Machinery Division	P.O. Box 19038 Green Bay, WI 54307–9038	(414) 494–4571
General Packaging Equipment Co.	6101 Westview Drive Houston, TX 77055	(713) 686–4331
Hayssen Manufacturing Co.	P.O. Box 571 Sheboygan, WI 53082–0571	(414) 458–2111
Mira-Pak, Inc.	7000 Ardmore Houston, TX	(713) 747–1100
Rovema (Pneumatic Scale Corp.)	65 Newport Avenue Quincy, MA 02171	(617) 328–6100
Triangle Package Machinery Co.	6655 West Diversey Avenue Chicago, IL 60635	(312) 889–0200

ADDITIONAL READING

Anon. 1987. Coffee line goes for distributed data logging. *Process Engineering* **68**(4): 19.

Anon. 1985. Challenges and changes coming for coffee cans. *Food Engineering* **57**(8): 54.

Anon. 1983. Metallized coffee pouch: ten years success (Elkin Coffee). *Food Engineering* **55**(12): 48.

Anon. 1983. Vacuum bag caters to consumers preference (Maxwell House). *Packaging* **28**(11): 12.

Anon. 1983. Pouch-packed instant coffee gains instant success in New Zealand (Nescafe). *Food Engineering* **55**(7): 39.

Anon. 1976. Flexibles: high barrier for fresher foods. *Modern Packaging* **49**(7): 21–24.

Buchner. 1975. New low pressure process for producing flexible vacuum and insert-gas packs. *Verpackungs Rundschau* **26**(5): 728–735.

Cameron, J. L. 1976. Paper laminations in flexible packaging. *Canadian Packaging* **29**(5): 38–39.

Hannigan, K. J. 1982. Coffee in an instant: it's in the bag. *Food Engineering* **54**(6): 47.

Hu, K. H. and Breyer, J. B. 1971/2. Pinhole resistance of flexibles. *Modern Packaging.* **44**(12): 46–48 and **45**(1): 47–49.

Neuton, J. R. 1977. Polyester as a metallized substitute. *Modern Packaging* **50**(1): 38–40.

Peters, J. W. 1975. Metallized film improves appearance and shelf life for institutional ground coffee. *Food product Development* **9**(7): 60–62.

Q 31: How Can Microwavable Foods Be Packaged?

A 31: It is estimated that more than 80% of U.S. households are equipped with microwave ovens. It is thus no wonder that more than $200 million of microwavable packaging material will be consumed in 1990. With all this success there are nevertheless some major problems left to be resolved.

HISTORY

The TV dinner preceded the advent of the microwave oven. The proliferation of single-occupancy homes and the two-breadwinner family created a need for ready-to-serve meals. The industry responded with frozen meals—in aluminum trays—ready to eat after a 25–45 minute oven heating. The idea of substituting plastics for aluminum had been suggested. Plastics have many interesting properties to recommend them: light weight, easy disposability (via incineration), better appearance (could be molded to resemble china), lower cost, and many others. There were, however, some disadvantages as well. Plastics are generally poor heat conductors and thus it would take longer to heat a meal in plastics than in metal trays. Furthermore, there was the danger of the plastic disintegrating under unspecified conditions. One must assume that some consumers will disregard instructions and place the frozen meal in an oven at a temperature much higher than the directions indicate, in order to have the meal ready in a shorter time. Others may follow the temperature directions but disregard the time restrictions and leave the meal in the oven while they shower, talk on the phone, etc. With the aluminum tray, such behavior may lead to a charred meal, but the package will remain intact. With a plastic tray, such disregard for directions may result in a meltdown of the tray. The consumer, no doubt, will blame the tray and not his misuse of it for the failure, in most instances.

While plastics were unsuccessful in making inroads into this market, a new composite, paperboard/foil, did offer an alternative. This material was especially suitable for cakes, cookies, and similar products.

At about the same time, the microwave oven made its appearance and changed the reconstituted meals market entirely.

THE MICROWAVE OVEN

One of the major problems facing the packer for this market is the lack of uniformity of equipment on the market. Practically all microwave ovens operate on a frequency of 2450 MHz. The wavelength at this frequency is 12.2 cm in air, but much smaller as the waves pass through materials. Thus the wavelength in water is almost a tenth that in air. There is the lack of uniformity of the magnetic field within each oven to contend with. This creates hot and cold spots within the oven and leads to uneven heating or cooking of the food. Packages often recommend the turning of the product 90° or 180° during the cooking cycle. This movement, or constant rotation on a turntable, is designed to distribute the heating effect more evenly throughout the product. However, the lack of uniformity between the various models on the market is staggering. The power output ranges from 300 to more than 750 watts, the oven size from 0.3 to 1.8 cu. ft. Some have just simple microwave, others also convection and still others have browning elements. It is thus understandable that the directions for microwave processing of food are sometimes ambiguous. The large variety of food products to be handled makes this task even more difficult. Frozen products complicate the task further. Water is highly absorptive and heats well, while ice is practically transparent to the microwaves and heats poorly.

Many microwave ovens are pulsing in order to improve heat distribution. This means that the power goes on and off at a preset cycle, 50% on and 50% off. However, the duration of each cycle may range from 1 second to 30 or even 60 seconds. The off period allows the hot spots to distribute the accumulated heat to other cooler areas. Nevertheless, it is quite common experience to find the center burnt and the edges very cool; conversely, especially in cooking frozen, high profile product, one may burn the outside while the center remains cool. These problems are both equipment and product related but are often blamed on the innocent package.

PACKAGING MATERIALS

At first glance, the material problem, if any, should be minimal. After all, hot food should be at about 170°F (76.6°C) and at this temperature almost any plastic should be able to survive. Theoretically, one should be able to utilize a boil-in-bag to heat any frozen product in a microwave oven. In practice this is not true. There are several reasons that preclude the use of a flexible pouch.

Multi-Course Package

Normally the meal consists of several courses, all on a compartmentalized tray. It is impossible to organize this meal in a pouch. One could have several individual pouches making up a dinner. However, this would defeat the "convenience" feature of such type of service. It should be mentioned that some of the fragile products could not survive in a bag.

Temperature/Pressure

In the pouch the pressure could build, threatening the integrity of the package and its contents. Since superheated steam bathes the package walls, the temperature of the pouch may exceed 300°F (148.9°C), causing serious damage to many plastic materials. Furthermore, the concern that the consumer will ignore instructions is ever present. To expedite the process, reconstitution may take place at higher power settings than specified or consumers may disregard instructions entirely and place the package in a convection oven rather than microwave. This is a possibility which must be reckoned with.

FDA

The package which was suitable for room temperature or freezer storage may not receive FDA sanction for elevated temperature use. The material in contact with hot foods may exude monomers, plasticizers, solvents, inks, resins, or low-molecular-weight polymers which would not migrate at lower temperatures. Thus a packaging material for microwave or conventional oven heating must not only survive the rigors of the temperature requirements, but must also pass the package/product interaction tests at the specified time/temperature cycles.

Rigids

The rigid package is preferred because of:

1. Compartmentalization. The meal can be presented in individual compartments for each course. With shielding, some courses (such as desserts) can be kept cool, while others are heated.

2. Serving tray. The plastic tray is not just a storage container in which the food is kept during shipment and while in the freezer. It is also a baking pan, a serving dish in which the hot food is brought to the table, and often the dinner plate from which it is eaten.
3. Disposability. The plastic tray can be incinerated or recycled. In either case it does not add to the mass of solid waste. The plastic also diminishes the danger of cuts one may contract from metal containers.

As mentioned above, ideally the package must withstand high temperatures due to the chance of local overheating. However, the potential misuse of a package (placing it in a convection oven) is a real possibility and thus the package should withstand temperatures in excess of 400°F (204°C).

C-PET

At this writing the most widely used plastic in the reconstituted food field is crystallized copolyester. For example, Kodar Termx® 6761 has received FDA approval for food packaging. It has a melting point of 545°F (285°C) and can thus be utilized for microwave or for conventional oven exposure.

Copolyester is not easy to process—but it can be done. It is not suitable for form-fill-seal equipment. It is not easily formed and must be allowed to crystallize in order to develop optimum heat resistance. The degree of crystallinity may range from 15 to 30%, but typically is targeted for 25%. The plastic sheet has an amorphous density of 1.195, but in 100% crystalline material the density increases to 1.265 during thermal crystallization. The lidding material is commonly composed of a heat-seal-coated polyester.

Polysulfone

This plastic can be utilized over a wide temperature range. It can withstand flash freezing as well as microwave or convection oven exposure, since it is serviceable from −150°F (−101°C) to +400°F (204°C). Even though a post-crystallization step is not required, this film is also not suitable for ffs equipment. The solution adopted in both these cases, with polysulfone or copolyester, calls for plastic trays to be formed on off-line molding equipment. Thereafter stacks of preformed trays are magazine fed onto ffs machines, where they are filled and sealed.

The wall thickness selected depends on package design, depth, product packaged and many other parameters. However, it was found that 14 mil (0.35 mm) was a workable minimum and about 40 mil (1 mm) an upper limit. Some years ago composites of polysulfone/polypropylene and polysulfone/HDPE were evaluated. The rationale for selecting these materials was manifold. The primary reason for the inclusion of a high-temperature-resistant polyolefin was cost containment. PP and HDPE are definitely less costly than either polysulfone or polyester. The polyolefin is also a preferred surface for lidding adhesion and furthermore, it offers improved barrier properties—especially moisture retention.

TPX

Polymethylpentene is claimed to melt at 455°F (235°C). It has excellent chemical resistance, good impact strength, and other interesting physical and mechanical properties. Some problems have been encountered with sheet formation. However, with proper predrying of the resin, good temperature control in the extrusion process, and attention to detail on the post-calendering, a very satisfactory sheet can be obtained.

TPX, too, has been evaluated as a composite in conjunction with higher melting PP or HDPE. The observation made under polysulfone apply here, as well. It has been noted that TPX has a slightly lower tolerance for elevated oven temperatures and should not be recommended for more than 350°F (159°C) exposure.

SUSCEPTORS

One of the prime shortcomings of microwave cooking is its inability to brown surfaces. The food item is cooked from the inside out, rather than from outside in, as in the traditional oven. In the old fashioned way, the outside is overdone and becomes crusty, crisp, and sometimes even burnt. In some foods, this crusting is most desirable. The microwave oven, heating the food from within, leaves the outside cool. The moisture from the center portion is driven toward the periphery, making the outside soggy. As some of this moisture evaporates the surface is consequently further cooled.

A number of approaches have been tried to remedy this problem. About 10% of microwave ovens have a hot-air convection feature. Thus some surface browning is achieved via hot air circulated through the microwave oven. Another approach involves special browning dishes and still another introduces susceptors for this purpose.

The susceptor may consist of a laminate composed of paper or board adhered to metallized polyester. The susceptor through reflection and/or absorption of microwave energy transfers some or all of this energy to the food surface, achieving the desired browning effect. The susceptor can become an integral part of the package or can function as inserts, strategically positioned around the food product. FDA has reservations regarding the potential food additive introduced by susceptors.

FOOD PRODUCTS

The items on a menu must be chosen with care. Not all are equally suitable for microwave reconstitution. Some foods will tolerate the vagaries of the microwave better than others. For example, it might be wise to include mashed potatoes rather than french fries. The latter will very likely turn out too soggy, while the gravy on the mashed potatoes will be hot and fragrant. A few thinner slices of meat might be preferable to a very thick slice. A hamburger in a bun may not be a good idea, unless they can be packaged separately (possibly in individual compartments of the same package). Each component of a meal must be evaluated for reconstitution not just be itself, but also in the presence of all the other items served on the same tray. Some products may have to be reformulated, others may have to be eliminated to create a harmonious serving.

ADDITIONAL READING

Anon. 1988. Ready meals convenience demand dual-ovenable performance. *Packaging* (UK) **59**(677): 24–26.

Gerling, J. E. 1990. Household microwave oven technology. *TAPPI Journal* **73**(3): 197–207.

Hirsch, A. and Spiegel, F. X. 1976. Bakable packaging films and packages. *Canadian Patent* No. 988830, May 11, 1976.

Huang, H. F. 1990. Specifying and measuring microwave food packaging materials. *TAPPI Journal* **73**(3): 215–218.

Huang, H. F. 1989. New product concepts in microwavable food packaging. *Microwave World* **8**(6): 5–7.

Hunter, B. T. 1988. Plastics in contact with food. *Consumers Research Magazine* **71**(11): 8–9.

Losenson, C. et al. 1990. Nonuniform heating of foods packaged in microwavable containers. *TAPPI Journal* **73**(3): 265–267.

Schiffmann, R. F. 1990. Microwave foods: basic design considerations. *TAPPI Journal* **73**(3): 209–212.

Wickersheim, K. A. 1990. On-line measurements in microwave ovens. *TAPPI Journal* **73**(3): 223–229.

Wood, A. S. 1987. There's a good reason for the action in polysulfone polymers. *Modern Plastics* **64**(2): 50–52.

Q 32: What Choices Are Available for Tamper-Resistant Closures?

A 32: The terms "tamperproof" or "tamper resistant" are incorrect. Some of the most sophisticated security systems have been unable to keep burglars from tampering with safes or otherwise secured premises. The best we can hope for is a "tamper evident" closure. Certainly, a package which may travel for hundreds or even thousands of miles from its point of manufacture to its sale to a consumer—this same package, which will have spent days on an open shelf, will have offered many opportunities to a tamperer to invade the package. We know for certain that many packages are smuggled out of the retail establishment to avoid payment for the product. It is also well known that many cases of product disappear in shipment. One accepts this pilferage as the cost of doing business. The consumer of course pays for the stolen items in increased product cost.

However, in recent years a number of incidents have caused the consumer to fear that the food (or drug) in the package has been altered and could possibly be dangerous. The most infamous of the tampering incidents is that of the Tylenol capsules. The drug content was mixed with cyanide and was thus transformed from a mild painkiller into a poisonous human killer.

The Tylenol incident taught us that packages can be invaded and the breach hidden from the average consumer. When Jaffa oranges were injected by terrorists with strychnine, we learned a similar lesson. The intrusion cannot be forestalled. The contaminant cannot be extracted or neutralized. However, the package can be designed to indicate that it was tampered with and the consumer is thus warned to avoid such food item.

We often hear at Halloween time of candy handed out to children by some misanthrope who introduced some contaminants to mete out retribution on the innocent children. In 1990 threats were made against M&M candies. A group with a specific political agenda claimed to have laced the candy with poison. The consumer has just one option to avoid serious harm—abstain from the suspect product. The manufacturer, however, faces high costs for recalls, loss of sales while product is removed from the retail shelf, and possible loss of market share since loyal customers are forced to seek alternate products and may not return after the crisis. Thus a prank can be ruinous to a business. No wonder, then, that tamperproof packaging is in high demand.

THE SIMPLE WRAP

The package should discourage tampering, but failing this it should disclose its violation in easily detectable terms. A piece of candy in a twist wrap is subject to tampering with impunity. No one can tell if the twist has been opened; a few grains of poison (or just a harmful substance) may have been placed on the candy and the twist wrap reclosed. Even slightly more complex packages can be invaded and all traces of the intrusion erased by an average person. The more complex the package, the greater the skill required to accomplish tampering without a trace.

The intrusion could be compared to a burglary. The amateur faced with a safe of the simplest type is stymied and gives up. The more experienced burglar will blow the safe, leaving a telltale destruction. However, the professional will manage to crack the most complex safe without explosives. He will open the door, extract the contents, and reclose the door, leaving no trace of mischief. It is this latter type of invasion of food packages which is fraught with serious consequences.

The simplest overwrap of Jaffa oranges might have dissuaded some terrorist from injecting poison into the fruit. More likely, however, the wrap would have been partially lifted, the poison injected, and the wrap replaced. The average consumer would have been unaware of the tampering.

MULTIPLE SEALS

A simple seal can be overcome. It is certainly possible, with a little skill, to open a package and reseal it. However, as the seals increase in number or complexity, the likelihood of achieving a perfect reclosure diminishes. Examples of this type are standard in many pharmaceutical packages. Tylenol, which has had its share of adverse publicity, has a shrink collar over a pressure cap. When both of these are broken there is an aluminum seal over the mouth of the bottle.

In many costlier snack items, we find an outer sealed plastic wrap, beneath which is a sealed box, inside of which there is a sealed bag of product. Unfortunately, the tamper scare has made this type of excessive packaging necessary. The extra cost is passed on to the consumer. Thus it is not the manufacturer or retailer who bears the wrath of the tamperer, but the innocent consumer who has to pay for it all. In addition, the convenience features in packaging are at risk. The fear of tampering has eliminated or at least curtailed many "easy open" packages, because they would offer the tamperer an opportunity to play his pranks. Some products such as capsules may have to be eliminated since they are prone to tampering.

PRINTING

Another approach to discourage tampering involves printed packages. The more complex the printing, the less likely that tamperer could duplicate it or could enter the package through a cut in the printed material and repair the cut to make it disappear. Attractively printed packages have been employed for a variety of reasons. Primarily, printing has been utilized as a sales tool. The attractive design entices the consumer to pick the product over a less striking competitive package. To discourage tampering the print must cover the entire package surface. If there are unprinted areas, then they become the ports of entry. The printing must be complex multicolor. Any simple print could possibly be duplicated by some amateur intending mischief. However, multicolor prints are much more difficult to forge.

TELLTALE SEALS

One of the victims of tampering is the easy-peel seal. In its place has come the telltale seal, a device that offers evidence of intrusion. Several systems are currently available. The destruct seal does not allow opening of the package without tearing or in some other fashion destroying the seal or the packaging material.

An alternative system utilizes a colored seal. If properly closed the seal has a deep blue or green (or any other) color. Once opened, this seal cannot be reclosed to achieve the same deep uniform color. Several other types of seal are available which disclose any opening and reclosing.

CONTROLLED ATMOSPHERE

A novel approach to protecting the integrity of a package may involve an acidic (CO_2) atmosphere and a pH indicator. As long as the package remains unopened, a little indicator spot or insert tab will be of one color—

green, for example. If the package is tampered with, the gas escapes and the pH within the package is altered. This in turn alters the color of the indicator from green to red and alerts the consumer to the tampering.

There are other systems either in use or on the drawing board. All are meant to alert the consumer to the intrusion rather than make such invasion of the package impossible.

ADDITIONAL READING

Alfred, C. 1987. Tampering major concern of food and drink industry. *Business Insurer* 21(10/18): 58.

Anon. 1987. New tamper-resistant package ideas. *Plastics Engineering* 43(6): 15.

Anon. 1986. Plastics steps up its role in fighting package tampering. *Modern Plastics* 63(9): 11–12.

Erickson, C. 1988. Bright ideas spur closure advances. *Packaging.* 33(12): 30–33.

Head, M. 1986. An ounce of prevention against tampering (food package). *Progressive Grocer* 65(2): 17–18

Herrin, D. T. 1987. The food industry's new villain (know how to counter the food tamperer). *Food Engineering* 59(2): 67–70.

Hunt, B. T. 1986. Tamper resistant packaging. *Consumer Research Magazine* 69(4): 8–9.

Kessler, F. 1986. Tremors from the Tylenol scare hit food companies. *Fortune.* 113(3/31): 59.

Larson, M. 1989. Tamper-evident in perspective. *Packaging* 34(5): 34–36.

Morgen, F. W. 1988. Tampered goods: Legal developments and marketing guidelines. *Journal of Marketing* 52(4): 86–96.

Muster, J. M. 1988. Tamper evidence and closure innovation. *Packaging* (UK) 59(677): 36–40.

Pick, G. 1986. Gerber's baby under stress. *Across the Board* 23(7/8): 9–13.

Skrzycki, C. 1986. Tampering with buyer confidence. *U.S. News and World Report* 100(3/3): 46–47.

Stillwell, E. J. and Rudolph, S. E. 1989. Strategies for foiling tamperers. *Packaging* 34(5): 39–41.

Q 33: *What Is the Meaning of Some of the Odd Vocabulary Common in the Packaging Language?*

A 33: Packaging has a set of unique vocabulary which is best explained by reference to this short glossary of terms:

ABHESION—The antonym of adhesion.

ABRASION—The damage caused by friction such as rubbing, scuffing or scratching.

ABSORPTION—The penetration of one substance into the mass of another.

ADHESION—The act or state of sticking together, uniting, or bonding.

ADSORPTION—A concentration of a substance at a surface or interface resulting from the attraction of molecules of the two substances.

AGE RESISTANCE—See *Shelf life.* The resistance to deterioration by oxygen and ozones in the air, and by heat, light and cold.

AMBIENT CONDITIONS—Surrounding environment. The temperature, humidity, and other conditions of the medium surrounding an object.

BANDS, BAGGY—A defect in roll stock materials; a slack lane, or a baggy section which shows up when the material is unwound and pulled taut. (See *Web bag* and *Web sag.*)

BARRIER—Preventing passage of oxygen, moisture or other gases (or odor) through the package.

BASIS WEIGHT—The basis weight for most packaging materials is calculated on a ream of 500 sheets of 24 × 36 inches, or in pounds per 3000 square feet of material.

BEAD—A thickened section at the edge of a roll of material.

BELLY—An excess fullness in cross section of the web, either in the center or near the sides. (See *Web bag* and *Web sag*.)

BLOCKING—The undesired adhesion of two or more plies of material, in roll, sheet or package form, to the extent that the surfaces become damaged or distorted.

BLOOM—The result of exudation of an ingredient from a product as visibly evidenced on the product or transparent package.

BLUSHING—Defects of bond quality of solvent adhesives or appearance of lacquer-type of coating resulting from the condensation of water from the ambient air.

BOND STRENGTH—A measure of the strength of a bond between two adherends.

BREAKING STRENGTH—The ability of a material to resist rupture by tension. A measure of the strength of paper, fabrics, films and other materials.

BRITTLE—Easily shattered or broken. In cellophane and paper usually the result of moisture loss.

BUCKLES—Defective roll formation usually caused by buildup of heavy internal pressures in wound roll. Buckles are deep, narrow folds in the roll, generally running parallel to the transverse direction.

BURSTING STRENGTH—The pressure required to rupture a specimen when it is tested in a specified instrument under specified conditions.

CALIPER—Thickness, generally expressed in thousandths of an inch or mils.

CAP—Controlled atmosphere packaging; either vacuum or gas replacing air in package.

CATALYST—A substance used to accelerate a reaction.

CHALKING—A dry, chalklike appearance or deposit on the surface of a material.

CHANNELING—A channellike delamination pattern in laminated materials. Also referred to as tunneling or worming.

CHEMICAL RESISTANCE—Ability of a material to retain utility and appearance following contact with chemical agents. Chemical-resistant properties include stain resistance, swelling resistance, moisture resistance, corrosion resistance, etc.

CLARITY—Transparency, freedom from haze.

CLING—A tendency of adjacent surfaces to adhere to each other, as in blocking, except that separation can be effected without damage to either surface.

COATING—The outer covering of a film or web. The film may be one-side coated or two-side coated.

COHESION—The state in which particles of a single substance are held together by primary or secondary valence forces.

COLD FLOW—A change in dimensions and/or shape of a material when subjected to external weight or pressure at room temperatures. In adhesives, the usually undesirable movement or flowout of the adhesive film. Most pressure-sensitive adhesives have this characteristic.

CONFORMABILITY—The ability of a packaging material to bend around sharp corners or projections without tearing or fracturing.

COPOLYMER—Polymer produced from a combination of two or more monomers.

COVERAGE—Dry weight of coating or adhesive used per unit of area.

CRATERING—Bare spots in a coating film which have the appearance of pock marks.

CRAZING—Fine cracks which may extend in a network on or under the surface of, or throughout, a film layer coating or adhesive or on surfaces of glazed materials such as glass, plastics, and painted or enameled surfaces. Also called checking.

CREEPAGE—The slight but continuous cumulative tendency of a color to drift out of register or position in the running direction.

CROCKING—Smudging, or rubbing off, of ink.

CROSS DIRECTION—The direction perpendicular to the *Machine direction.*

CURL—An undesirable condition caused by uneven stresses. Curl is most prevalent in coated or laminated materials where the components have differing physical properties.

DEAD FOLD—A hand- or machine-made fold which will remain in position without sealing or pressure, such as on a soft foil.

DELAMINATION—Partial or complete separation or splitting, usually caused by lack of adequate or sufficient adhesion in laminated or plied materials.

DWELL TIME—Time of exposure to a specified condition.

ELASTOMER—Any rubberlike substance or polymer.

ELONGATION—Lenghtwise stretch of a material usually expressed as percentage of its original length.

EQUILIBRIUM—Point at which a substance neither gains nor losses a stated property.

EXUDATION—Migration of an ingredient in a material or product to the surface.

FADING—The loss of color strength on exposure to light, heat, or other agents.

FIBER-TEAR—In a paper lamination, the tear of fiber as opposed to separation of the laminant when the assembly is pulled apart.

FILM—Unsupported, basically organic, nonfibrous, flexible material of a thickness not exceeding 0.010 inch. Such material in excess of 0.010 in thickness is usually called sheeting.

FISH-EYES—Particles of undissolved or extraneous material in a film or coating resembling the eye of a fish.

FLAT—In packaging materials, the degree of freedom from curl, or other distortion.

FLEXIBILITY—The property of a material which permits its being bent or twisted without breaking.

FLEXING STRENGTH—The ability of a sheet or film to withstand breakage by folding. Flexing strength may be measured by a test to determine the number of folds required to cause failure.

FFS—See *Form/fill/seal*.

FORM/FILL/SEAL—Process or machinery forming material into a container, filling same with edible product, and finally sealing the container shut, all in a continuous operation.

GAS TRANSMISSION RATE—A measure of the permeability of a packaging material to gases by measuring the movement of a gas through the film under specified conditions.

GAUGE—An instrument for exact measuring. See also *Caliper*.

GHOSTING—In printing, transfer of the design image from printed to unprinted areas which may occur between any two adjacent surfaces without actual ink transfer.

HALO—An unwanted line surrounding a printed image caused by excessive pressure.

HAZE—A cloudy or foggy appearance in a normally clear transparent material.

HDPE—High-density polyethylene. See *Polyethylene*.

HEAT RESISTANCE—The ability to withstand the effects of exposure to high temperature.

HERMETIC—Airtight or impervious to air or fluids.

HOMOGENEOUS—Of the same composition or construction throughout.

HOT-MELT LAMINATING—The use of either 100% resins, or wax fortified with resins to the extent that they behave as a true thermoplastic material.

HUMIDIFY—To moisten or dampen. To cause the atmosphere surrounding a product to contain moisture.

HYDROPHILIC—Having a strong affinity for water. Refers to colloids which swell in water and are not easily coagulated.

HYDROPHOBIC—Lacking affinity for water. Opposite of *Hydrophilic*.

HYGROSCOPIC—Having the property of readily absorbing moisture from the atmosphere.

IMPACT STRENGTH—Resistance of a material or product to shocks such as from dropping and hard blows.

KISS—The lightest possible impression which will transfer the film of ink from the transfer roller to the plate and from the plate to the web.

LAMINATE—(1) (noun) A product made by bonding together two or more layers of material or materials. (2) (verb) To unite layers of materials with adhesives.

LAMINATION—Process of plying layers of stock to a given thickness.

LDPE—Low-density polyethylene. See *Polyethylene.*

LEAKERS—Defective packages which permit the undesirable escape of any portion of the contents by sifting, exudation, permeation, etc. or a loss of vacuum by an evacuated package.

LOOSE WIND—Rewind made loosely with insufficient tension, often resulting in telescoping.

LOW-TEMPERATURE FLEXIBILITY—The pliability of a material, such as a plastic, at low temperature.

MAP—Modified atmosphere packaging. Containing gases other than air.

MIL—A unit of linear measurement, equivalent to 0.001 inch.

MOISTURE CONTENT—The percentage of water in a finished material.

MOLDABILITY—Ability of a material to lend itself to bending, folding, creasing, kneading, or otherwise conforming to the contour of an object or mold and to retain its acquired shape.

MONOMER—A chemical combination of molecules corresponding to the individual units of a polymer. It is capable of being incorporated (polymerized) into polymers.

MVT RATE—Moisture vapor transmission rate. Incorrectly used. See *Water Vapor Transmission Rate (WVTR).*

NEUTRAL—Absence of acid or alkaline activity on a material. A substance having a pH of 7.

NONFLAMMABLE (non-inflammable)—Will not support combustion.

NONFOGGING FILM—Film which does not become cloudy from moisture condensation caused by temperature and humidity changes.

NYLON—A synthetic resin which belongs to the polyamide family.

ODOR, RESIDUAL—Any foreign odor which is retained by a material or product after a portion has passed off or dissipated. In packaging materials, residual odors are commonly caused by the solvents in inks, coatings, and/or adhesives which were not completely driven off during the manufacturing process.

OFF COLOR—In films, slit roll edge displays extreme variation in color or shade; off-color film is not necessarily defective in any respect and will almost always perform properly.

OFF GAUGE—Not within specified dimensions of thickness.

OPACITY—Resistance of a material or body to transmission of light.

OPAQUE—Not permitting the passage of light.

PE—See *Polyethylene*

PET—Polyethyleneterephthalate, a very durable polyester resin.

PEELING BOND—The amount of force necessary to separate two sheets actually determines the strength of the adhesive bond.

PERMEABILITY—Ability to permit passage of gases or liquids. See *Barrier* and *Gas Transmission Rate.*

pH—A numerical representation of the acidity or alkalinity of an aqueous solution. 7 is neutral; below 7 down to 1 indicates increased acidity; above 7 up to 14 indicates increased alkalinity.

PINHOLING—(1) In printing, the failure of an ink to form a completely continuous film, resulting in fine holes in otherwise solidly printed areas. (2) In metallic foil, very small holes through the thickness of the foil which normally are not evident until the foil is held against a strong light source.

PLASTICITY—That property of a body by virtue of which it tends to retain its deformation after reduction of the deforming stress.

PLASTICIZER—An agent or compound which is added to materials, to impart either softness or flexibility, the lack of which will cause the material to be rigid or brittle.

PLASTICIZER MIGRATION—Undesired movement of the plasticizer to the surface of a material or from one material to another.

PLY—One of the layers in a lamination.

POLYAMIDE—Polymers containing amide groups, for example, Nylon, Versamide resins, etc.

POLYESTER—Reaction product of a polyol with a polyacid. Resin is very stable and may yield films of considerable packaging interest.

POLYETHYLENE—A synthetic resin of high molecular weight resulting from the polymerization of ethylene gas under pressure.

POLYMER—A compound formed by the linking of single and identical molecules having functional groups that permit their combination to proceed to higher molecular weights under suitable conditions.

POLYMERIZATION—A chemical reaction in which the molecules of a monomer are linked together to form large molecules whose weight is a multiple of that of the original substance.

POLYPROPYLENE—A synthetic resin of high molecular weight, resulting from the polymerization of propylene gas.

POLYSTYRENE—A thermoplastic material derived from the polymerization of styrene.

POLYVINYL ALCOHOL—A family of resins produced by hydrolysis of polyvinyl acetate.

POLYVINYL CHLORIDE—A family of resins produced by the polymerization of vinyle chloride.

POLYVINYLIDENE CHLORIDE—See *PVDC*.

PP—See *Polypropylene*.

POROSITY—The quality or state of being permeable.

PSI—Pounds per square inch, gauge pressure rather than absolute pressure. Used in stating air-pressure, bursting strength, etc.

PVC—See *Polyvinyl chloride* and *Vinyl*.

PVDC (Polyvinylidene chloride)—An adhesive available in solvent solution or water emulsion form whose dried film is thermoplastic and exhibits good barrier resistance to water, grease, oil, and many vapors.

REAM—The unit of quantitative measure used in the marketing of paper. The ream most widely used for packaging materials consists of 500 sheets, each 24″ × 36″.

REGISTER—To have one part positioned accurately with respect to another. In package printing, refers to accuracy of position to secure correct alignment of color-to-color areas, or of design-to-scores shown on the die sheet, or the correct placement of the design on the printed areas or items.

REGISTRATION—In overwrapping operations, the placing of the design in proper relation to the package faces.

RESINS—Natural or synthetic complex organic substances with no sharp melting point which in solvent solution form the binder portion of a flexographic ink.

RETARDERS—Solvents added to ink to slow the evaporation rate.

SCUFF—To rub or abrade.

SEALABILITY—That property of a material which renders it capable of being sealed.

SEAL STRENGTH—The measured force required to separate a seal.

SELF-CURING DRY BOND ADHESIVE—A two-part solvent system based on urethane resins with a crosslinking resin that will complete its chemical reaction at a temperature low enough to permit rewinding the web during the process. Many adhesive suppliers now offer such adhesive systems.

SHEAR—An action resulting from applied forces which causes or tends to cause two continuous parts of a body to slide relatively to each other in a direction parallel to their plane of contact.

SHELF LIFE—The length of time that a container, or a product in a container, will remain in a salable condition.

SKIP—To miss or jump over a spot.

SLIP—The degree of slipperiness of a material or its comparative lack of drag, which is essential for ease of handling and good machineability.

SOFTENING POINT—Temperature at which plastic material will start to deform with no externally applied load.

SOLVENT—The medium used to dissolve a substance.

SOLVENT COATING—A type of coating applied in liquid form, which dries by evaporation.

SOLVENT RETENTION—In packaging materials, the undesirable condition which occurs when the solvent in inks, coating, and/or adhesives are not completely dissipated.

SPLICE, INCORRECT—Splices improperly made using the wrong color tape, or other error for intended end use.

SPLICE, WEAK—Splices improperly made; insufficient strength for end use.

SPOILAGE—Material spoiled in process of manufacture; waste caused by mistakes; errors in judgment, or faulty processing; deterioration of food by microbiological action.

STABILITY—The quality of durability or constancy of properties; applied particularly to emulsions, to certain foods and chemical compounds or to dimensions of a material.

STABLE—Having constant properties; not easily altered by changing temperature, humidity, aging, mechanical manipulation, etc.

STATIC ELECTRICITY—Stationary charges of electricity which sometimes develop in bodies during handling or in machine operation; may cause undesired attraction of film to rollers, flat surfaces, etc.

STATIC ELIMINATOR—A device for removing static electricity.

STORAGE LIFE—The period of time during which a packaged product can be stored under specific conditions and remain suitable for use. Sometimes called shelf life.

STRENGTH—The mechanical properties of a material which permit it to withstand parting or distortion under the application of force. Strength properties include toughness, tensile strength, flexural strength, tear strength, flexibility, etc.

STRETCH—Extensibility of materials under tension.

STRIATION—A streaky pattern of parallel grooves in the print.

SUBSTRATE—Any material to which adhesives, inks, or coating are applied, printed, or extruded. Substrates can include film, foil, paper, board, etc.

TACK—In adhesives, commonly regarded as the stickiness. The amount of tack is estimated as the pull-resistance when attempting to separate the adherends while the adhesive still exhibits viscous or plastic flow, the separation being effected without failure or deformation occurring in the adherend surroundings.

TACK, INITIAL—The tack of an adhesive immediately after application.

TEAR STRENGTH—Resistance of a material to tearing.

TELESCOPING—Transverse slipping of successive winds of a roll of material so that the edge is conical rather than flat.

TENSILE STRENGTH—The resistance of material to longitudinal tension stress.

TRANSPARENT—Transmitting rays of light so that objects can be seen through the material.

TENSION—The stress caused by a force operating to extend, stretch, or pull apart.

TEST—Perform an operation designed to determine a specific quality of the material.

TEST, DROP—A test in which filled containers are dropped from controlled heights by means of a special device to insure uniformity of drop. Used as a laboratory means of simulating the rough handling a package will receive during shipment.

THERMOPLASTIC—Capable of being resoftened by heat.

THERMOSET—A material which hardens when heated and does not soften when reheated.

TOLERANCE—Permissible maximum deviation from specified dimensions or qualities.

TRANSLUCENT—Permitting passage of light but diffusing it so that objects cannot be seen clearly.

VAPOR TRANSMISSION—The passage of vapor (usually water vapor) through a material. The properties of a packaging material permitting the passage of vapor.

VINYL—Informal generic term for any of the vinyl resins, or for film or other products made from them.

VISCOSITY—Resistance to flow.

VOLATILE—Easily passing from a liquid into a gaseous state, subject to rapid evaporation. Having a high vapor-pressure at room temperature.

WATERPROOF—Extreme resistance to damage or deterioration by water in liquid form.

WATER VAPOR PERMEABILITY—The ability of a material to permit transmission of water vapor. Test of permeability of packages or packaging materials are done under fixed conditions of time, temperature and water vapor pressure-differential between the two sides of a material or the inside and outside of the container.

WATER VAPOR TRANSMISSION RATE (WVTR)—Formerly called moisture vapor transmission rate. The actual rate of water vapor transmission used to compare water vapor barrier, wrapping, or container materials. Usually expressed in grams of water passing through 100 square inches of material in 24 hours at a specified temperature and relative humidity.

WEB—The roll of paper, foil, film, or other flexible material as it moves through the machine in the process of being formed, or in the process of being converted, printed, etc.

WEB BAG—The condition that exists when the center section of the web droops or sags under tension while the sides are taut. The opposite of *Web sag.*

WEB GUIDE—A device which keeps the web traveling straight through the press.

WEB SAG—The condition which exists when the sides of the web sag or droop under tension while the center is taut. The opposite of *Web bag.*

WET LAMINATING—The combining of two plies of material by use of a wet adhesive whose solvent element must seep into one of the plies to set up adhesion. Sodium silicate, dextrines, and latexes are typical wet adhesives.

WET STRENGTH—A measure of the physical strength properties of paper when saturated with water, expressed in terms of wet tensile-strength, etc.

YIELD—The coverable area per unit weight of materials such as films, foils, etc., usually expressed as square inches per pound.

YIELD POINT—That point beyond which the stresses in a material will cause a permanent deformation.

Non nobis, D., non nobis;
sed nomini tuo da gloriam
Psalms 115:1

Index